TO FOSTER THE SPIRIT OF PROFESSIONALISM

History of American Science and Technology Series

General Editor, LESTER D. STEPHENS

*The Eagle's Nest: Natural History and American Ideas,
1812–1842*
by Charlotte M. Porter

*Nathaniel Southgate Shaler and the Culture of American
Science*
by David N. Livingstone

*Henry William Ravenel, 1814–1887: South Carolina
Scientist in the Civil War Era*
by Tamara Miner Haygood

*Granville Sharp Pattison: Anatomist and
Antagonist, 1781–1851*
by Frederick L. M. Pattison

*Making Medical Doctors: Science and Medicine at
Vanderbilt since Flexner*
by Timothy C. Jacobson

*U.S. Coast Survey vs. Naval Hydrographic Office: A
19th-Century Rivalry in Science and Politics*
by Thomas G. Manning

*George William Featherstonhaugh: The First
U.S. Government Geologist*
by Edmund Berkeley and Dorothy Smith Berkeley

Homicidal Insanity, 1800–1985
by Janet Colaizzi

*Curators and Culture: The Museum Movement in
America, 1740–1870*
by Joel J. Orosz

*Enter the Physician: The Transformation of Domestic
Medicine, 1760–1860*
by Lamar Riley Murphy

*To Foster the Spirit of Professionalism: Southern Scientists
and State Academies of Science*
by Nancy Smith Midgette

TO FOSTER
THE SPIRIT OF
PROFESSIONALISM
Southern Scientists and
State Academies of Science

NANCY SMITH MIDGETTE

Ga Diss, 1984

The University of Alabama Press

Tuscaloosa and London

Library of Congress Cataloging-in-Publication Data

Midgette, Nancy Smith.
　To foster the spirit of professionalism : Southern
scientists and state academies of science / Nancy Smith
Midgette.
　　p.　cm. — (History of American science and
technology series.)
　Includes bibliographical references and index.
　ISBN 0-8173-0549-1 (alk. paper)
　　1. Science—Southern States—Societies, etc.—
History.　2. Science—Southern States—History.
3. Research institutes—Southern States—History.
4. Research—Southern States—History.
　I. Title.　II. Series.
　Q11.M58　1991
　506.075—dc20　　　　　　　　　　　　　91-7765

British Library Cataloguing-in-Publication Data available

Contents

Acknowledgments

The debts that I have incurred as this study has come to fruition are many. Lester D. Stephens first kindled my interest in the origins and development of the southern state academies of science; his friendship and encouragement sustained me through the process of research and writing. The interest of Malcolm M. MacDonald in this study, long before it was ready for the University of Alabama Press, also kept me in front of the word processor when I could have as easily quit.

Countless archivists and librarians worked long and hard on my behalf as I combed through both the collected records of the state academies of science and papers of various persons active in those organizations. I must thank especially the staffs of the University of Georgia, the Southern Historical Collection at the University of North Carolina, Chapel Hill; the Special Collections Department of the J. Y. Joyner Library, East Carolina University; the W. S. Hoole Special Collections Library, University of Alabama; the North Carolina State Archives; and the Special Collections Department, South Caroliniana Library, University of South Carolina. Additionally, I must express my gratitude to Beveley S. Powers of the Auburn University Libraries; Glenn L. McMullen of the University Libraries, Virginia Polytechnic Institute and State University; Alex Sartwell of the Alabama Geological Survey; and David Sowell of the Tennessee State Library and Archives.

Many members of the various state academies of science have also provided great assistance. Among them are Libby Workman, Wintfred L. Smith, James X. Corgan, George Palmer, Emmett B. Carmichael, George Rogers, and Boyd Harshbarger.

A number of colleagues offered valuable support and suggestions as my work proceeded: William F. Holmes, Joseph R. Berrigan, Numan V. Bartley, Thomas G. Dyer, and the readers for the University of Alabama Press. The staff members of the University of Alabama Press have been most gracious. Craig Noll's masterful and painstaking editorial skills saved me from a number of errors. This study is the stronger for their efforts. I alone am responsible for any errors that remain.

My colleagues at Elon College have always known when to be sympathetic, and when to be encouraging. My students have helped me retain my sense of humor when life seemed less than amusing.

Finally, I must express my deep gratitude to my parents, William R. and Helen Hayes Smith, who fostered my inquiring mind; to William C. Harris, who directed me toward the historical profession; and to my husband Charlie, without whom I could not have enjoyed this great adventure.

Brief portions of this work appeared in a considerably different form in journals: "In Search of Professional Identity: Southern Scientists, 1883–1940," *The Journal of Southern History* 54 (November 1988): 597–622; "The Alabama Academy of Science and the Maturing of a Profession," *The Alabama Review* (April 1984): 98–123; and "Vanguard of a New Generation: The Elisha Mitchell Scientific Society and the Scientific Profession in the South," *Journal of the Elisha Mitchell Scientific Society* 100 (Summer 1984): 65–73. Permission to use this material has been granted by the journals.

Organization: A Vital Component of Professionalism

Organizations have become, in the twentieth century, a vital component of professionalism. Historians, sociologists, engineers, medical doctors, physicists—all relate to one another through organizations. Thomas L. Haskell, one of several scholars who have analyzed the evolution of such societies, astutely noted in his study of the American Social Science Association that scrutinizing the life cycle of these organizations should not be "an end in itself but a point of entry into a larger subject."[1] In fact, the story of the development of professional societies mirrors the development of the professions themselves, from the point where men and women with similar training and interests first recognized a commonality among themselves, to today's sophisticated career ladder. This study, based on the foregoing premise, utilizes the southern state academies of science as a vehicle for exploring both the growth of the scientific profession in the South and the relationship between southern scientists and their colleagues nationwide.

Any discussion of the growth of professional science, whatever the region of the nation, must be grounded in an understanding of exactly what the term "professional" implies. A number of historians of science have addressed this issue.[2] As Nathan Reingold has noted, defining professionalism is a difficult task, in part because of the term's many connotations.[3] For centuries society has associated professionalism with physicians, clergy, and lawyers, implying not only a certain educational level and specific job-related skills but also a degree of responsibility to humankind. The term has also carried an elitist connotation; society admired these persons and expected them to maintain a high standard of moral and ethical conduct.

By the middle of the nineteenth century, an increasing number of men had acquired formal training in the sciences, largely in European universities, and had secured employment that required the use of their specialized knowledge, either as professors in American colleges or as scientists with the United States Coastal Survey or the various state geological surveys. Considering themselves professionals, they were anxious to distinguish themselves from persons whom Reingold terms "cultivators," learned individuals with some "specific knowledge in the sciences" who did not, however, use their scientific expertise as a means of employment.[4]

One avenue to distinction was restricted association. While scientists had historically mingled with the "cultivators" in general learned societies such as the American Philosophical Society of Philadelphia and the American Academy of Arts and Sciences of Boston, by the mid-1800s they began to alter this pattern. In 1840 geologists organized the American Association of Geologists and Naturalists (AAGN), which soon attracted the attention of scientists in other disciplines. In 1847 the geologists voted to broaden the AAGN into the multidisciplinary American Association for the Advancement of Science (AAAS).[5]

According to Sally Gregory Kohlstedt, the founders of the AAAS were determined to create a "viable organization . . . with the potential to coordinate scientific inquiry and to establish science as a true and visible profession in the United States."[6] It would appear, then, that research—advancing knowledge as well as diffusing it—had become an integral element of the scientists' concept of professionalism. The AAAS began auspiciously; membership grew, as did participation at meetings. The organization's growth, however, contained the seeds of discord. The constitution permitted anyone with an interest in science to join the society, and this open membership policy caused a number of persons to complain that little had been accomplished in the effort to distinguish professional scientists.[7]

Elements against distinction
— Democratic groups
— Scarcity of journals
— extreme elitism ? NAS

Nor did the National Academy of Sciences (NAS), organized in 1863 by some of the disgruntled AAAS members, serve their purposes any better. They hoped that this organization, initially limited to fifty of the nation's most prominent scientists, would (1) identify them as the professional elite, (2) establish them as advisers to the government on any and all matters requiring scientific expertise, and (3) enable them to entice research support from Congress. However, the organization encountered enmity and distrust from its inception, since many competent scientists had been omitted from the initial fifty members. Then too, its fortunes shifted with the ever-changing political wind, and the coveted governmental support was usually not forthcoming. Over time the membership ceiling rose, and although election to membership continued to be an honor, it was not essential to a career in the sciences.[8] *Dupree*

Even more frustrating to aspiring professional scientists than the democratic nature of the AAAS and the elitism of the NAS was the scarcity of adequate journals. Specialists disdained the popular periodicals and found that the import tax on European scientific treatises priced these publications beyond their usually modest means. They were, of course, familiar with the *American Journal of Science*, begun in 1818 by Benjamin Silliman, a Yale professor of chemistry, mineralogy, and geology. However, this interdisciplinary journal had its limitations, too, as young scholars discovered. In 1873 Henry Augustus Rowland, then a physics instructor at Rensselaer Polytechnic Institute but soon to be professor of physics at the Johns Hopkins University and a renowned scientist, submitted an article to editor James Dwight Dana. Following customary procedure, Dana sought an expert opinion from the physicists at Yale University. Later admitting that they did not understand Rowland's mathematics, they suggested to Dana that he decline to publish the article. Soon thereafter, James Clerk Maxwell published it in the British *Philosophical Magazine*.[9] In short, the young, well-educated disciplinary specialists pushing the forefront of knowledge often found themselves at the mercy of well-meaning but mediocre colleagues. *How often ?*

In fact, scholars such as Rowland, most of whom had received their graduate education in European universities, wondered if

America would ever produce first-rate scientists, meaning those dedicated to research. In an address to the AAAS in 1883, he denigrated most of America's "322 so-called colleges and universities," pointing out that only 17 of them employed a faculty of twenty or more persons and that a mere 8 had endowments in excess of $1 million.[10] While he noted that both teaching and commercial invention were necessary and respected vocations, professional scientists required ample opportunity for "the true pursuit of their science"—basic research. Positioned in a few select universities, these professionals, Rowland maintained, could devote their energies to expanding knowledge and, at the same time, instruct and serve as an example for students who would "see before them this high and noble life."[11] No doubt these comments ruffled a number of collegial feathers, given Rowland's well-equipped laboratory at the Johns Hopkins University.

In 1980 historian Daniel Kevles, in conjunction with Jeffrey Sturchio and P. Thomas Carroll, offered a somewhat more balanced view of American science in the late nineteenth century. They surmised that in 1880 about 3,300 people "to some degree used science in their employment." About 500 of these people could be considered "serious publishing researchers." Another 2,000 persons were cultivators, in Nathan Reingold's sense of the term. While, conclude the authors, most of the "serious researchers" probably agreed with Rowland, it is erroneous to generalize about a very diverse population from the comments of one person. Certain branches of scientific investigation, including physics, might have been weak; but others, such as the earth sciences and astronomy, boasted some first-rate scientists. Moreover, Rowland, directing his remarks primarily to fellow academicians, appeared to overlook the scientists employed in industrial laboratories, assaying offices, public health, agricultural experiment stations, the United States Geological Survey, and the United States Weather Service. In fact, one of the strengths of late nineteenth-century American science was its "pluralism of institutional identity. . . . No single aim or patron governed the American scientific enterprise."[12]

Such diversity dictated that financial support for scientific investigation was spread among a variety of institutions—colleges and universities, industry, state and federal governments—each with their own agendas. For the most part employers expected scientists to serve specific purposes, which did not often include undirected, basic research. In institutions of higher education, where one might most likely find such activity, scientists usually faced such demanding undergraduate teaching and laboratory instruction loads that they had little time or inclination to pursue independent research. Except for Johns Hopkins, founded in 1876 for the purpose of research and graduate training, most colleges and universities rewarded teaching over research and supplied pedagogical rather than investigative laboratories.[13]

Despite these odds, scientific disciplinary specialists remained determined to establish criteria that would define their professional status. To this end they moved beyond the AAAS and organized discipline-based societies such as the American Chemical Society (1876), the Geological Society of America (1888), and the American Physical Society (1899). Although membership requirements varied, all of these organizations retained more control over who joined than did the AAAS; at the same time, they were more open to young scholars than the NAS. Led by an elite cadre, these organizations formulated professional career parameters that included a graduate degree, original research and publication, employment utilizing specialized scientific knowledge, and membership in one or more scientific organizations.

Far more than just an external scaffold for career development, this concept of professionalism as analyzed by Burton Bledstein became an internalized "set of learned values and habitual responses."[14] Persons wishing to enter the ranks of professional science accepted and met these criteria. Coincidentally, as graduate schools in the United States expanded, they supported this concept of professionalism and infused it into their students. Young graduates, many with masters' degrees and some with doctoral degrees, entered the work force with the sense that a professional scientist must labor to increase knowledge as well as use and disseminate it. Thus the concept of professionalism as it evolved among late nineteenth-century scientists established the goals for disciplinary aspirants.

The scientific organizations that materialized during the last quarter of the nineteenth century served as a stimulus for this professionalism by providing a network of communication that included opportunities for people to meet together to discuss their work and share research results, and in some cases a publication outlet. Scientists in northeastern urban areas, where colleges abounded and graduate schools and industrial research laboratories emerged with some regularity, found that participation in these organizations, usually centered in a nearby city, was a relatively easy process. Scientists in more remote areas, however, experienced frustration because of their relative isolation and the difficulty of traveling to distant meeting sites.

Such was the plight of scientists in the American South, a traditionally agrarian region that could boast of but few urban areas and supported only a handful of creditable colleges. Until fairly recently, the reputation of the American South concerning intellectual pursuit in general and scientific investigation in particular was not good. Clement Eaton, in his study of the Old (antebellum) South, awarded honorable mention to a handful of the region's scientists before declaring that "in the Southern states there existed a special influence that militated against the development of the scientific attitude, namely, the subtle and pervasive effect of slavery." Since then, scholars have moderated this intellectual epitaph by delving into the lives of individual southern scientists and asking new questions that expose the extent of the region's scientific interests and accomplishments. They conclude that a number of well-educated scientists lived in the region but that their intellectual growth was constricted by a lack of urban centers, inadequate libraries, and limited contact with persons sharing similar interests. Moreover, employment opportunities for scientists existed almost exclusively in the region's few educational institutions, dictating that many of them could pursue their interest only as an avocation while earning a living in another manner.[15]

Scientific pursuit in the postbellum South has received remarkably little attention from historians, perhaps because of the continuing fascination with Reconstruction politics and race relations. To be sure, professional scientists in the South between 1865 and 1940 are difficult to identify and enumerate. Most institutions of higher education, both state sponsored and private, suffered severe underfunding, which translated into low salaries, high work loads, and minimum budgets for books, equipment, and travel. In short, the region offered little to entice well-trained scientists to move south. Consequently, while most colleges and universities employed men and women to teach the scientific disciplines, only a minority of these persons held a doctoral degree, and even fewer were recognized as scholars outside their local communities. Those who did choose employment in the South found their professional associations limited to the colleagues in their own institutions and contact via mail with distant friends and organizational headquarters.

Yet it would be a mistake to dismiss the post–Civil War South as a scientific desert. These educators were keenly aware of professionalizing forces within their disciplines, and as time passed, they struggled to gain recognition within this crystallizing scientific community. Administratively powerless within their institutions, they coped as best they could with poorly funded departments and set their sights on creating opportunities for professional advancement. Tentatively at first, southern scientists organized local societies that most often revolved around a state university and attracted little attention outside the immediate community. Then, beginning in 1902, they followed the lead of their colleagues in other states and created statewide, interdisciplinary academies of science that they hoped would fill the void left by their inability to participate actively in national organizations. While the primary goal of these societies was to provide an opportunity for scientists to meet together, they attempted to publish creditable journals and to generate incentives to research activity.

Prior to World War II, the southern state academies of science were only moderately successful in meeting the professional needs of their members. They did hold annual meetings where members could mingle with colleagues and present papers. Most of them published a journal, but usually these publications contained little more than minutes of the annual meetings, officers' reports, and abstracts of papers presented at the meetings. The organizations were even less successful in stimulating research, for two key ingredients—funding and reduced teaching loads—did not materialize until after World War II.

Despite their poor performance with regard to their original purpose, these state academies were not failures. In the pre–World War II South they represented the only source of professional contact for many scientists. They also served as a channel of communication whereby the handful of scientists who did participate in national organizations could share ideas and concerns with their local colleagues. Although difficult to prove, the professional ideals espoused by these local organizations no doubt stimulated a number of science instructors to pursue a graduate degree and perhaps some research, thus upgrading the quality of scientific instruction and investigation in the region. Finally, the academies facilitated community involvement for scientists regarding such issues as conservation of the environment and the quality of science education in primary and secondary schools.

Following World War II, the South's phenomenal economic growth generated an academic environment far removed from the impecunious and isolated one in which southern scientists had so long labored. Rapidly expanding colleges and universities hired more and better-trained scientists, funded modern research laboratories, and channeled considerable resources into graduate programs. Modern transportation enabled these men and women to participate in national professional organizations; consequently, many southern scientists, especially those who joined the ranks of the profession during this heady era, saw little reason to support the state academies of science. It appeared that, having served their purpose at a significant moment, these societies had become an anachronism.

A few stalwart academy leaders, though, thought otherwise and sought ways to attract more members. During the 1950s and the 1960s they chose to focus academy attention on the service component of professionalism, championing two causes that could better be addressed at the regional than the national level—the quality of secondary science education and conservation of the environment. Aided by federal funds and sometimes by private bequests, the academies not only began junior academies of science but also sponsored science fair competitions and scholarship awards designed to entice talented students to pursue scientific studies. Additionally, many of the academies monitored potential abuses of the local environment, taking appropriate action when a delicate area seemed threatened.

Meanwhile, they did not abandon their functions of the earlier era. State academies of science continued to sponsor annual meetings and to publish journals, even knowing that national societies and their publications would be the outlet of first choice for most scientists. By the mid-1950s, academy leaders accepted this situation but maintained that the regional organizations could still fill a vital need by offering young scholars an opportunity to present their first papers and by providing a forum whereby more-established scholars could claim priority for research that would eventually be published elsewhere.

The southern state academies of science were born of a particular need, and changing times produced a new environment and new needs. Their significance can be compared with that of organizations such as the American Philosophical Society, the Boston Society of Natural History, the AAAS, and the NAS. All of these societies represented crucial links in the chain of scientific professionalism. Their demise or alteration of purpose mirrors the evolving nature of the scientific profession in America. Likewise, the history of the southern state academies of science exemplifies the struggle of southern scientists to bring the region to a par with the rest of the nation and to earn for themselves the respect that they felt they deserved. Consequently, the decline in importance of the southern state academies of science is not indicative of intellectual stagnation in the region but, rather, signifies that these men and women have found it possible to become an active part of the national professional scientific community.

2 Scientists in the Postbellum South

February 19, 1865, dawned crisp and clear, a "fine bracing morning" recalled Joseph LeConte much later. This professor of geology and chemistry at South Carolina College had left Columbia four days earlier along with his brother John (a member of the same faculty), John's son, Major Allen J. Green (a Confederate army officer), and a number of servants and wagons in a desperate attempt to keep their own property and that of the Confederate Nitre Bureau out of the hands of William T. Sherman, whose forces were rapidly closing on the city. Just as the small traveling party had settled down to breakfast, a warning shout pierced the air: Union soldiers were upon them. Men scattered; Joseph LeConte rushed toward the wagons that contained his family's jewelry and silver, along with his manuscripts. Too late. A small band of blue-clad soldiers pilfered the valuables, and as LeConte watched helplessly from his nearby hiding place, they set fire to the wagons. In a matter of moments his life's work was reduced to ashes. The troops discovered John, his son, and some of the servants and marched off with their prisoners in tow. Joseph, fearful for the safety of his brother and nephew, trudged back to Columbia to find the rest of his family safe but much of the city a charred ruin. Putting the pieces of his life back together would be an arduous process.[1]

While most southern scientists did not face such a harrowing ordeal, they found themselves in dire straits at the end of the Civil War. Political instability and severe financial hardships retarded the development of the single most important institution for professional scientists—colleges and universities. Yet despite such adversity a handful of these persons eventually won enviable reputations and assumed an active role in the national scientific community that blossomed during the last quarter of the nineteenth century. Rather than deserting the South for greener academic pastures, many of them devoted considerable effort to improving the quality of scientific investigation and instruction in the region. They continued to teach undergraduate students, inaugurated graduate instruction, participated in national professional associations, and remained active researchers.

They also conveyed to both their southern colleagues and their students the importance of accepting and meeting the criteria of professionalism then becoming such an integral aspect of a scientific career. In addition to promoting these ideals within their own institutions, they organized local scientific societies that they hoped would establish a sense of camaraderie and foster the spirit of inquiry so much a part of the national scientific community. While these late nineteenth-century organizations did not live up to the aspirations of their founders, they served as forerunners of societies that would be of considerably greater significance for southern scientists—state academies of science. These early organizations represent a crucial component in the development of the modern spirit of southern scientific inquiry.

Scientific Inquiry in a Rebuilding South

The postbellum South contained fewer scientists than did other regions of the nation. Of 166 scientists listed in the *Dictionary of American Biography* who entered upon their careers between 1861 and 1876, only 8.3 percent lived and worked in the South.[2] The destructive effects of the Civil War impeded the development of the South's professional scientists far more than it did those in the North. Supported primarily by educational institutions, scientists found less than fertile ground in Dixie in which to plant their careers. First the Confederate government and then Union troops had confiscated campus buildings for their immediate housing needs, paying little heed to the valuable books and equipment lost or destroyed in the process. A number of colleges closed for a period of time, setting faculty members adrift. Salaries for professors who remained in their positions were often reduced and sometimes withheld completely.

The unstable political atmosphere of the former Confederacy further hampered the development of scientific inquiry in the region. Southern politicians jockeyed for position as the national balance of power shifted from President Andrew Johnson and his relatively mild plan of Reconstruction to a more radical Congress. The resultant military governments imposed on ten states of the South, as well as the first state administrations elected under federal supervision, provided a less than auspicious environment for the recovery of state educational institutions.

Southern scientists responded to this postwar environment in a variety of ways. Joseph LeConte, having already lost his manuscripts, grew steadily more despondent over the Reconstruction government in South Carolina and the potential effect upon the state's university. The problem was not financial, for the state legislature had appropriated operating funds and even included a salary increase for faculty members. Weighing more heavily on his mind was the possibility of black domination of the state government and perhaps the administration of the university as well, for he accepted the racial prejudices of the day. Although his job and that of his brother, who had been released from captivity, in all probability were not in jeopardy, the LeContes sought employment outside the South. Throughout 1867 and 1868 the brothers wrote to colleagues around the nation and eventually received appointments at the newly created University of California, an institution they served for the remainder of their lives.[3]

Scientists at the University of North Carolina suffered a more severe jolt to their careers than did the LeConte brothers, for in 1868 the college closed for seven years. Charles Phillips, the premier faculty member at the time, found employment as professor of mathematics at Davidson College, near Charlotte; in order to make financial ends meet, he also served as pastor of a nearby church. John Kimberly took up farming near Asheville. While both of these men returned to the university when it reopened in 1875, others did not. Manuel Fetter finished his career at various private academies in the state, and Fordyce Mitchell Hubbard went to St. John's College in Manlius, New York.[4]

Nowhere was the future more bleak than in Tuscaloosa, Alabama, where Union troops had burned all of the state university buildings. President Landon C. Garland valiantly tried to continue instruction, but the appearance of only one student in the autumn of 1865 shattered his optimism. Forced to weather the storms of economic hardship and the vagaries of contemporary politics, the University of Alabama did not resume full operation until 1871.[5]

Throughout the 1870s southern colleges struggled just to reclaim (or rebuild) their physical plants and to assemble faculties. Meanwhile, colleges in the Northeast and Midwest began an exciting metamorphosis as they expanded enrollments, broadened curricula, and inaugurated graduate studies in many fields. The development of professional and graduate schools that trained young men and women for specific careers challenged the assumption that a college education was the preserve of an elite social class, or that it should be grounded in a study of the classics. A growing number of these institutions also accepted a public responsibility by offering services such as agricultural extension programs to the communities in which they were located.

Southern educators were not oblivious to this shift in philosophy concerning higher education. Andrew Adgate Lipscomb, president of the University of Georgia from 1860 until 1874, wanted among other things a scientific curriculum on the order of those at Harvard and Yale. Realizing that the trustees would never accept his plan and that it probably would be financially impossible anyway, he settled for a broadened curriculum that de-emphasized the classics, introduced a limited course of scientific study, and allowed juniors and seniors the privilege of electing some of their course work.[6]

Other southern state universities also initiated considerable curricular revision. The University of North Carolina *Alumni Quarterly* of 1894, the first issue to appear after the Civil War, heralded the changes. The chemistry department, announced the *Quarterly*, offered, in addition to its customary general course, qualitative and quantitative chemical analysis and organic, industrial, agricultural, theoretical, and historical chemistry. Other departments similarly expanded their specialized instruction. While courses at the freshman and sophomore levels continued to emphasize "thorough drill and culture in such studies as constitute the foundation of a liberal education," the last two years offered students the opportunity for "advanced work in special lines, for original thought and investigation, and for special preparation for chosen professions or lines of business."[7]

The University of Virginia, whose scientific curriculum before the Civil War had been "almost rudimentary," followed suit. By 1881 the Charlottesville institution boasted "schools" of Natural Philosophy; Mathematics; Natural History and Geology; Agriculture, Zoology and Botany; Practical Astronomy; and Analytical, Industrial, and Agricultural Chemistry.[8] While most of these "schools" were in reality little more than one professor teaching several related courses, their presence is indicative of some specialization among the science faculty and a growing emphasis on scientific instruction. Although prior to the turn of the twentieth century only the state universities of Virginia and North Carolina expanded to the point that they awarded the doctoral degree, others throughout the region offered masters' programs in the sciences and strengthened scientific courses in the undergraduate curriculum.

In addition to expanding state universities, southern higher education made some further headway when the Georgia Institute of Technology opened in Atlanta in 1888. Southern boosters heralded it as an institution that would emulate the Massachusetts Institute of Technology and train a new generation of specialized professionals. Tech, though, operated on the "shop culture" approach rather than that of the "school culture." As one study has noted, it "placed greater stress on practical shop work and produced graduates who could work as machinists or as shop foremen, but who were not well prepared for engineering analysis or original research." MIT, on the other hand, "stressed higher mathematics, theoretical science, and original research."[9] Faced as well with a consistently niggardly legislature, Georgia Tech was unable to attract top quality faculty and did not blossom into a truly modern institution until the 1940s.

The development of Georgia Tech mirrored the situation in colleges and universities throughout the South. Financial stringency dictated that faculty must bear heavy teaching loads; resources for experimental laboratories and travel to national professional meetings simply were not available. The scarcity of major endowed universities in the South at this time further accounts for much of the region's poor showing in producing both scientists and original research. For the most part, southern institutions funded the expansion of their scientific instruction with the provisions of the 1862 Morrill Federal Land Grant Act, which awarded federal lands (and the potential income from their sale) to states that would initiate programs of study in agriculture and the "mechanical arts." This legislation meshed well with the pragmatic economic orientation of the South and, over time, produced knowledgeable farmers and important agricultural research and extension services. It did little, however, to encourage southern colleges and universities to sponsor a broad base of research and instruction; consequently many scientific disciplines languished.

Nonetheless, southern institutions did manage to attract some able faculty. Among the ten faculty members appointed in 1871 to the rebuilding University of Alabama was Eugene Allen Smith, professor of mineralogy and geology, who soon became Alabama's leading scientist and earned a prominent national reputation. Smith was born on October 27, 1841, in Washington (Autauga County), Alabama, the son of a physician, Samuel Parrish Smith, and Adelaide Julia Allen Smith, a direct descendant of Governor William Bradford of the Plymouth colony. The Allens first arrived in Alabama in 1823 from Connecticut; Smith's grandfather opened a tannery. They remained in the state for seven years, returning to Connecticut in 1830 to provide for the education of their only daughter, Smith's mother. In 1836 the family returned to Alabama, and shortly thereafter the young woman married Samuel Smith. The couple made their first home in Washington, but soon after their son's birth moved to Prattville, Alabama. Smith received his early schooling at a local private academy, but at age fifteen journeyed to Philadelphia, Pennsylvania, to enroll in the renowned Central High School, perhaps following the pattern set by his mother in returning east to complete his education.[10]

In 1860, at age nineteen, Smith returned home and enrolled as a junior at the University of Alabama. He received his A.B. degree in 1862 and then joined the Confederate army. Although he enlisted as a private, his company elected him second lieutenant (a common practice at the time, left over from bygone militia days). In December 1862 Confederate president Jefferson Davis appointed Smith instructor of tactics at the University of Alabama, where he remained for the duration of the war. His closest encounter with action came when Union troops advanced on the university in 1865; the cadets, being no match for seasoned veterans, chose not to make what would have been a suicide stand and watched helplessly as the school was destroyed by fire.[11]

At some point in his life Smith developed an interest in the study of science. In October 1865 he put the ruined Confederacy behind him and traveled to Germany to continue his education, a common path for Americans who sought specialized scientific training then unavailable in the United States. Smith spent one semester in Berlin, one at Göttingen, and then two years at the University of Heidelberg, where he received the Ph.D. degree. Upon returning to the United States, he secured a position at the University of Mississippi as instructor of chemistry and found himself a colleague of Eugene Woldemar Hilgard, a renowned geologist who had previously served the Smithsonian Institution and would subsequently teach at the University of Michigan before accepting the position of professor of agriculture and director of the Agricultural Experiment Station at the University of California in 1875. Smith remained in Oxford until 1871, working closely with Hilgard on the Mississippi Geological Survey as well as teaching his classes. Then he learned of his appointment as professor of chemistry and mineralogy at his alma mater.[12]

Unlike John and Joseph LeConte, who chose to leave the South, Smith was elated when he heard of his appointment to the University of Alabama faculty. Of course, his situation was somewhat different from that of the LeContes. Whereas the two brothers had already established their scientific reputations and sought to continue work begun earlier, Smith was only entering the profession, and his call to Alabama presented an opportunity to return home. Furthermore, by 1871 Alabama had already passed through the worst of the political turmoil that the LeContes sought so desperately to escape. To be sure, evidence of Reconstruction remained in the state, for in 1873 when Smith accepted the position of state geologist (in addition to his professorship), he stipulated that it carry no salary above his compensation from the university. Otherwise, he wrote to a colleague, "some radical ignoramus would work himself into the Survey for the sake of the spoils."[13]

Before his death in 1927 Smith had earned a considerable reputation and had assumed leadership positions in the national scientific community. He maintained membership in the American Association for the Advancement of Science, the Geological Society of America (GSA), and the American Institute of Mining Engineers. He served the AAAS as a sectional vice-president in 1904, and the GSA as vice-president in 1906 and as president in 1913. His 116 publications included a variety of reports for the Alabama Geological Survey and the United States Geological Survey as well as articles in the *Proceedings of the American Association for the Advancement of Science, American Journal of Science, Science, American Geologist, Journal of Geology*, and *Bulletin of the Geological Society of America.*[14]

Another prominent southern scientist of the late nineteenth century was Francis Preston Venable (1856–1934) of the University of North Carolina. Venable was born in the family home, Longwood, Prince Edward County, Virginia. His mother's family, the McDowells, were prominent Virginians; his grandfather, James McDowell, had served a term as Virginia's governor. Venable's father, Charles Scott Venable, was a young mathematician serving on the faculty of the University of Georgia at the time of his son's birth. Shortly thereafter, he accepted a position at the University of South Carolina, and the family moved to Columbia.

When the Civil War broke out, Charles Venable joined the Confederate army, and mother and child returned to Longwood. In 1865 the elder Venable received an appointment as professor of mathematics at the University of Virginia.[15]

As a youngster Frank enjoyed the run of the campus at Charlottesville, Virginia; he displayed an affinity for both classical languages and chemistry and could often be found in the laboratory of chemistry professor John W. Mallet. A special friendship developed between the two. Venable assisted Mallet with his research, which included determining the atomic weight of aluminum, and Mallet, on the night that Venable's mother died, worked with the young man as he prepared oxygen for her.[16]

Following his graduation from the University of Virginia in 1877, Venable taught at the University High School in New Orleans (run by his cousin) for a year and then returned to Charlottesville to complete work for a master's degree. Mallet then advised him to continue his education at a university in Germany. In October 1879 Venable sailed to France and then traveled by train to Bonn; he studied under August von Kekule for a number of months and then followed his mentor, Mallet, to the University of Göttingen.[17]

Not long after settling in at Göttingen, Frank received a letter from his father summoning him home—he had just been elected professor of chemistry at the University of North Carolina! Frank's seeming good fortune was actually the result of an exchange of letters between UNC president Kemp Battle and the elder Venable. The two men had been friends for years, and in the summer of 1880 Battle, desperately in need of a chemistry professor, wrote inquiring letters to a number of colleagues, including Venable. The Virginia mathematician recommended his son, and when Battle's first choice, Charles W. Dabney, who had already received his Ph.D. from Göttingen, seemed unavailable, Battle turned to Frank Venable. The young Venable hurried home from Europe (returning to Göttingen briefly in the summer of 1881 to complete his degree) and in October 1880 assumed his duties in Chapel Hill.[18]

Francis Preston Venable served the University of North Carolina for his entire career—fifty years as professor of chemistry and fourteen years (1900–1914) as president of the institution. He presided over the beginnings of the doctoral program and occupied the university's first endowed chair, the Mary Ann Smith Professorship. He authored several textbooks and countless articles. In 1892 he assisted John Motley Morehead, a former student and then chief chemist of Willson Aluminum Company near present-day Eden, North Carolina, by identifying an unknown by-product as calcium carbide. Venable's published paper on this effort saved Thomas R. Willson's patent rights (and the future Union Carbide Company) through a number of lawsuits. In his later years, Venable's best-known research concerned the atomic weight of zirconium, deposits of which are located in western North Carolina. He had long thought that the accepted atomic weight was too low. Utilizing a new method for determining atomic weights developed by Nobel Prize winner Theodore Richards of Harvard, he combined his own research with the discovery by the Danish chemist von Hevesy that zirconium in its natural state contains traces of another element, hafnium, to prove his thesis. In 1924 the American Chemical Society published his monograph on this subject.[19]

Certainly scientists of the stature of Smith and Venable were exceptions rather than the rule in the late nineteenth-century South. The majority of their colleagues had not earned a doctoral degree, nor were they engaged to any extent in research activities. While not oblivious to the currents of professionalization, they found that pursuit of a doctoral degree, original research, and participation in national organizations were limited by heavy teaching loads and inadequate financial support. Moreover, their lack of contact with scientific colleagues engendered a sense of isolation. While dedicated both as students of their disciplines and as instructors, most southern scientists remained anonymous academicians.

The Elisha Mitchell Scientific Society

In an effort to offset their sense of isolation, scientists within a single community would organize local societies. Thus was born the LeConte Scientific Society in Columbia, South Carolina; the

Joseph LeConte Society in Knoxville, Tennessee; and the University Scientific Society in Athens, Georgia. These societies made no attempt to emulate national organizations but, rather, served as focal points for academic scientists at the University of South Carolina, the University of Tennessee, and the University of Georgia respectively. In a more specialized vein, organizations like the Raleigh (North Carolina) Biological Club attracted not only faculty from local academic institutions but nonacademic specialists as well.

In 1883 the small science faculty at the University of North Carolina, led by Francis Preston Venable, decided to organize a society more akin to national professional organizations than to the aforementioned local societies. Certainly the Chapel Hill campus was fertile ground for such an organization, for relative to its size and isolated southern location, the university had attracted well-educated and aspiring young scholars. Joining Venable was Joshua Walker Gore (1852–1908), who taught engineering and physics, and Joseph Austin Holmes (1859–1915), professor of geology and mineralogy. The school had also recently inaugurated a doctoral program, and the faculty anxiously sought to instill in the students the professional spirit that they had absorbed in European and northeastern universities.

Gore, like Venable a native Virginian, arrived in Chapel Hill in 1882 as professor of natural philosophy and engineering. He had graduated from Richmond College, received a bachelor's degree in civil engineering from the University of Virginia, and studied math and physics for two years at Johns Hopkins University. Prior to accepting the position with the University of North Carolina, Gore taught physics, astronomy, and chemistry at Southwestern Baptist University in Jackson, Tennessee, and then served briefly in the mathematics department at the University of Virginia. During his early tenure at UNC Gore was responsible for the lighting, heat, and water plants as well as for his courses of instruction. In 1902 Gore became dean of the School of Mining, and in 1904 dean of the Department of Applied Science. Although not nationally recognized as a scholar, Gore published five papers during his career and earned the respect of his North Carolina colleagues.

He was recommended for the presidencies of both the University of North Carolina and North Carolina A & M College (later North Carolina State University) but withdrew his name from consideration, perhaps because of declining health. Gore died of tuberculosis in 1908.[20]

Joseph Austin Holmes, born in Laurens, South Carolina, graduated from Cornell University in 1881 with a bachelor of science degree and in the fall of that year became professor of geology and natural history at the University of North Carolina. Intermittently he continued his education, receiving a doctoral degree from the University of Pittsburgh. Like Venable, he belonged to several national professional organizations, including the AAAS, and was a charter member of the Geological Society of America. In 1893 Holmes became state geologist for North Carolina. He continued to teach at the university until 1906, when he left the state at the behest of President Theodore Roosevelt to organize the technological branch of the United States Geological Survey. His concern over the waste of mineral resources soon broadened into a campaign for mine safety; thereafter he supervised research into the causes of mine explosions and traveled extensively to promote safety. In 1910 Holmes became the first director of the United States Bureau of Mines. Five years later severe tuberculosis forced his retirement.[21]

Venable and Holmes became fast friends shortly after their arrival in Chapel Hill. Still in their early twenties, the two men prior to marriage briefly "kept bachelor's quarters" together in a house on the university campus. Years later, Venable recalled that the arrangement "had its inconveniences, since his [Holmes's] passion for collecting made him gather a miscellaneous assemblage of insects and reptiles, some of which had an uncanny habit of wandering down the stairs in the night to visit my quarters."[22]

Also vivid in Venable's memory was the environment in which the young scholars found themselves.

We were both just boys, he a year or two the younger, full of enthusiasm and energy for the big work which we realized was before us, and yet we were sadly hampered by the lack of almost everything in the way of books and equipment to which we had been accustomed. For the Uni-

versity of that day was struggling to rise from the desolating effect of
Reconstruction Times, received no support from the State which it had
so faithfully served in former years, and, with little money and only a
few devoted friends had to meet bitter opposition and misunderstanding
on every side.

Venable continued by pointing to the lack of even the most basic
instructional tools. "There was no University library in those
days, or, at least, the books gathered in a preceding age were kept
practically locked up. So we spent our own small savings in
gathering a few books around us and in providing any special apparatus."[23]

Venable's gloomy picture was no exaggeration. Other than tuition fees, the university's main source of income was interest from
monies received from the sale of federal lands granted by the Morrill Act, resulting in a total annual institutional income of approximately twenty thousand dollars.[24] It is little wonder that these
southern scientists could not attend national professional meetings—they could not even obtain adequate library materials and
laboratory equipment. Suffering from their isolation, Venable and
Holmes finally concluded that "for our own salvation, if for no
higher reason, we must gather some kindred souls about us, and
by the elbow-touch ward off the deadening effect that isolation
was bound to have upon our scientific work."[25]

The concern of these young scholars reflects their growing sense
of professionalism. Even under the most desirable of circum-
stances, with adequate equipment, funding, and relief from a full
teaching load, scientific research was a demanding and often frustrating task. In the South, it bordered on the impossible. Yet Venable and Holmes, the former fresh from stimulating study in the
universities of Germany and the latter then in the process of obtaining the doctoral degree, decided that they "could not afford to
vegetate and lose the productive years of their lives."[26] While they
exercised no control over the university's financial impediments,
they could mount a concerted effort to increase their contact with
fellow scientists.

Thus on September 24, 1883, the two young men invited a
number of colleagues to their home to consider the formation of a

scientific society. Attending the meeting, in addition to Venable and Holmes, were professor of mathematics Ralph H. Graves; president of the university Kemp P. Battle; professor of anatomy Thomas W. Harris; professor of law John Manning (Venable's future father-in-law); William B. Phillips, Venable's first doctoral student, then nearing completion of his degree; and Emile Alexander De Schweinitz, a graduate student just beginning his work under Venable and serving as an assistant in chemistry. They chose to name their fledgling organization in honor of Elisha Mitchell, one of the university's premier antebellum scientists.[27]

In advance of this meeting, Venable, Holmes, Graves, and Gore circulated a letter among their colleagues throughout the state outlining the proposed organization and soliciting support. "More thorough scientific training and increased interest in scientific work is a clearly felt want in the State," they noted. "The building up of a true spirit of scientific research cannot be effected all at once, but it is believed that the formation of a Scientific Society, thus bringing about a union of strength and effort, will aid greatly in increasing the zeal of those already at work and arousing the interest of others."[28]

The authors denoted five objectives of the society: (1) to cultivate a general interest in natural history and other scientific subjects; (2) to encourage individuals at work in the sciences "who isolated as they are at the present, with no one of congenial pursuits to turn to for advice or encouragement, are apt to become disheartened or indifferent and give up their work"; (3) to increase the knowledge of the state and its resources; (4) to foster smaller, local societies; and (5) to maintain a specimen collection. Members were to pay a small fee, in return for which they would receive the proposed annual publication, containing reports of original research by the members. Finally, the society hoped to sponsor a series of public lectures, "popularizing science as far as possible."[29]

Reflecting upon the moment forty years later, Venable wrote that "nothing short of the optimism and audacity of youth could have proposed and carried out such a plan." Serious as the young men were about the organization, their enthusiasm bubbled forth with a sense of humor as well. Venable remembered that they "camouflaged their identity and borrowed prestige for their bold adventure by adopting the names Hump, Rump, Mike, and the Blank. These pseudonyms stood respectively for no lesser personages than Sir Humphrey Davy, Count Rumford, Michael Faraday, and an humble unknown."[30] No doubt this was an amusement that they enjoyed among themselves, for the circular announcing the organizational meeting of the society revealed their true identities.

The constitution of the Mitchell Society reflected the professionalism of its organizers. While at first glance the provisions for maintaining a specimen collection and for popular public lectures would seem to be a throwback to the antebellum learned societies and their goal of disseminating knowledge to the general populace, such was not the case. No doubt the desire for a specimen collection, which never materialized, reflected the lack of one at the university. And the apparent anachronism of public lectures in fact indicated a certain amount of professional security. Three decades earlier, organizers of the National Academy of Sciences had shunned this form of public responsibility as unprofessional. Although in part their distaste for public lectures stemmed from their irritation at being forced into such a role to earn a living, they also sought to remove themselves from the environment of the general learned society and doubted the wisdom of disseminating specialized scientific knowledge to the public. By the 1880s, however, scientists in the Mitchell Society were sufficiently comfortable with their image as professionals to reach out and inform the public of the rudiments of their work. From that time on, scientific societies, while organized as a result of professional concerns, increasingly accepted a role of public responsibility that by the middle of the twentieth century extended far beyond dissemination of knowledge.

Despite their concern for public awareness, the founders of the Elisha Mitchell Scientific Society chose to emulate the new professional organizations by instituting rather stringent membership requirements. Only persons either actively involved in some form of scientific work or directly affiliated with the university would be eligible to join. In addition to honorary members (those persons recommended by the council and elected by a three-fourths vote of the membership, presumably for scientific achievements), the constitution provided for regular and associate members. Regular members were nominated by at least three current regular members and elected by a three-fourths vote. Undergraduate students at the university were tendered membership in the society as associates, with all privileges of regular members except that of voting.[31] The society encouraged students to join, for as Venable stated in 1884, one of the aims of the society was to present "treatises on scientific subjects" with "the hope of interesting and training up a number of young scientific workers."[32]

The monthly Saturday meetings were designed to appeal to faculty and students alike, to be both informative and entertaining. Of particular importance to the founders was the opportunity to stay abreast of the latest research, as would be the case were they able to attend national meetings. The constitution provided that "scientific papers and contributions to the Journal shall be read and discussed, and lectures delivered on subjects of general interest for the benefit and improvement of Associate and other members; also reports on progress in the various branches of Science, and brief biographies of distinguished scientific men." Moreover, an annual journal "shall contain original contributions from the members of the Society on Scientific subjects" which had first passed the scrutiny of the executive committee.[33]

While the founders sought membership from throughout the state, they realized that transportation difficulties and other commitments would often prohibit people from attending the monthly Saturday meetings. In order to facilitate the smooth transaction of business, therefore, the constitution stipulated that at least one of the two vice-presidents and the secretary be Chapel Hill residents. The resident vice-president could then preside in the president's absence, and the secretary, responsible for recording the proceedings, would be available on a regular basis.[34]

With mechanics in place, the Mitchell Society formally convened on November 10, 1883. Unfortunately records of attendance, if kept at all, have not survived, and so it is impossible to know how many persons were present on that day. Dr. Charles

Phillips, professor of mathematics emeritus, appropriately opened the program with a biographical sketch of Elisha Mitchell. How pleased Phillips must have been to see eager and well-trained young faculty as a part of the institution he had served for so long. The years of Reconstruction, when the university had closed and he had temporarily left Chapel Hill, must have seemed vindicated as he took his seat to hear President Venable's inaugural address.

Venable's remarks left little doubt that he was speaking mainly to his professional peers rather than to a general audience. "The primary object of this as of other scientific societies," he proclaimed, "is the advancement of our knowledge of Nature and Nature's laws." Researchers, Venable maintained, must understand that the smallest areas of investigation contribute to the whole body of knowledge. "Advance in knowledge now is only to be made by unceasing, careful and thorough labor." Armies of workers would be required to further our knowledge of the world. No longer could scientists afford to hold back their work until they felt it to be "rounded and complete," for that day most likely would never come. Rather, he said, "we publish our thoughts . . . and often our crude notes that others may help by their suggestions or criticisms."[35] Here was the epitome of the spirit of professional scientific research.

Venable hoped that the Mitchell Society would serve aspiring young scientists and more established professionals alike. The journal, he insisted, "will interest the workers, bring about a friendly emulation among them and give them a chance to show the world that they are neither dead nor sleeping." Moreover, the monthly meetings "will bring the members together and rub off the rust that is apt to gather on minds not brought in contact with other minds." Finally, he hoped that the meetings would be made "attractive and instructive by lectures, scientific papers and reports on progress in science," lifting student and instructor alike from the "ruts" and the "hum drum of prescribed courses" and "giving freshness and life to what they are apt to look upon as task work."[36]

Venable concluded his remarks with the well-worn "love for our country" (meaning the South) motive for pursuing scientific investigation in the face of formidable odds.

Southern
identity

In all our Southern land so blessed by nature as it is, few voices are raised above the din of traffic and the turmoil of bread-winning, to elevate the mind and teach the world—few eyes are opened, few hands outstretched to win from nature the secrets she so jealously guards. . . . We are falling behind in the world's progress, and whatever of civilization and of knowledge we possess we tamely accept from others. This must not be. It shall not be if we truly love our country. The stigma must be removed. . . . It has been the dream of my childhood, it has been the hope of my youth, it is the purpose of my manhood to give my labor and my life to the cause of elevating my beloved South. . . . I do so gladly welcome the formation of this Society. It is at least an entering wedge into the log of apathy and ignorance which we have to split.[37]

During the academic year 1883–84 the Mitchell Society convened seven times. The program for the second meeting, held in December, contained only five papers, but the May program listed twenty-two papers, reflecting the growing popularity of the society. Although many of these papers were read "by title only" (a common practice in societies until the 1930s, whereby only a brief abstract was actually presented, primarily to alert scholars to research in progress), their presence nonetheless indicates that at least some scholars were quick to take advantage of the forum thus provided. Of the sixty-seven papers presented that first year, fourteen represented the work of undergraduate associate members; the founding members of the society were responsible for most of the rest: Venable (presented fifteen), Holmes (eight), Graves (three), Gore (four), Phillips (five), DeSchweinitz (three). These papers, brief by modern standards, reported original research and observations, offered progress summaries on a variety of topics, and reviewed recent literature. Each program usually contained one paper in a more popular vein as well. At the April meeting, J. A. D. Stephenson, a Statesville resident, reported on a recent tornado; the following month, Charles Phillips recounted the history of the UNC observatory.[38]

By the end of the first year, the treasurer reported a balance of $134.42 (the journal had not yet been printed), and the membership roster listed 82 regular members and 66 student associates.[39] An explanation of the large student membership is problematic, since the total student body of the University of North Carolina in 1883 numbered only 205 persons. Apparently, students were en-

couraged to support the new society, whether or not they had
chosen to pursue scientific studies. Such would not be surprising,
because every regular professor at the university, regardless of field
of interest, joined the organization (three temporary instructors of
English, Greek, and Latin did not join). At this time the university
community was still quite small, and although some of the faculty
members perhaps had an interest in the proceedings of the society,
many of them probably joined out of respect for their colleagues.

This large initial membership plus the encouraging appearance
of twenty-nine members, either in person or by proxy, in May
1884 to elect officers for the coming year gave the founders great
hope for the future of the organization. Washington Caruthers
Kerr, a native of Guilford County, North Carolina, was elected
president. Kerr received his Ph.D. from Harvard in 1857, taught
for a time at Davidson College, and served as North Carolina's
state geologist before accepting a position with the United States
Coast Survey in 1882. William J. Martin, professor of chemistry and
geology at Davidson College, was elected vice-president. The remaining officers represented UNC: Gore, resident vice-president;
Venable, secretary and treasurer; Graves, Holmes, and Phillips, the
executive committee.[40]

Despite this encouraging start, the core of the organization soon
expressed disappointment with the turnout for the monthly meetings. Of course, they should not have expected more than a handful, because members living more than a few miles from Chapel
Hill faced a long, tedious journey in order to attend. Furthermore,
the meetings, structured to meet the needs of professionally minded
members, were not likely to catch the popular imagination. During
its second year the society, hoping to attract the interest of the local
townspeople, inaugurated a series of public lectures. Topics included
"The Domestic Life of the Romans," "Alchemists and Alchemy,"
"History and Objects of Geodesy," and "Biography of Dr. [Moses
Ashley] Curtis." As with the regular monthly meetings of the society, no attendance figures survive, but Vice-President Gore noted
that "members of the University and citizens of the town attested
their appreciation of the opportunity for instruction these lectures
afforded by their presence and attention."[41] Such lectures occurred
periodically over the next decade.

Decline in
members

While the annual officers' reports indicate an active, thriving society, the membership roster tells of a more sedate organization. By 1888 the roll contained the names of twelve honorary members and ten corresponding members, most of whom lived outside the state and therefore did not participate in the monthly activities. A total of sixty-six regular members, forty-one of whom lived less than sixty miles from Chapel Hill, and eleven associate members reflects a considerable decline from the first year. Undoubtedly many alumni of the university who joined the society during its early months allowed their memberships to lapse a short time later, especially if they lived any great distance from the university community. Likewise, many of the students who joined at the outset probably soon lost interest. Others graduated, further depleting the ranks of the associate members.

For a decade the Mitchell Society continued much as it had begun. University of North Carolina faculty dominated both the monthly meetings and the officers' ranks, although residents of Raleigh, Durham (Trinity College), Wake Forest (Wake Forest College), and Charlotte (Davidson College) occasionally appeared on the programs and among the lists of officers. The society sponsored several public lectures each year and published its journal, which contained officers' reports, program listings, and several papers. However, by 1892 long-distance participation had declined, and the membership, recognizing that the Mitchell Society had become in fact a local scientific society, voted to limit itself to university faculty and students. Consequently, the *Journal of the Elisha Mitchell Scientific Society* in reality became a university bulletin containing reports of research from the various scientific departments on campus. After 1893 the *Journal* eliminated reports of the society's proceedings and did not include regular officers' reports, minutes of meetings, membership lists, or even an indication of when meetings were held and the nature of the programs.

While the Mitchell Society continued to meet at least sporadically, it certainly did not become what its founders had envisioned. No doubt Venable was saddened by this turn of events. In a philosophical if somewhat resigned manner, he maintained in 1894 that the society continued to fulfill an important function on the Chapel Hill campus, because its meetings provided a good change of pace for students and professors, served as a stimulus for research, and offered at least a modest outlet for publication.[42]

President Kemp Battle, in his history of the University of North Carolina, agreed with Venable. He noted the declining membership, and hence the treasury, of the society. The trustees of the university, however, "seeing the value of the annual publication," voted to support the *Journal* with a one-hundred-dollar annual appropriation. Battle echoed his approval, maintaining that the society "has proved a valuable aid" to the science faculty and students of the university, "chiefly in stimulating the professors and advanced students to original work." He noted with pride that about 20 percent of the papers submitted to the society had been the work of students, "a small modicum of which would have been executed if there had not been the stimulus of publication."[43]

The Mitchell Society represented a transitional phase in the development of the scientific profession in the South. Some of its original goals, in particular the idea of maintaining a specimen collection and the hope for establishing branch organizations throughout the state, proved impractical. The society did, however, meet many of the immediate needs of its members. It offered them the camaraderie so necessary to productive scientific research and afforded them a modest publication outlet. By 1888 the society had established a journal exchange program with 129 other scientific and learned societies across the nation and throughout the world, thereby giving its own publication a wide audience and at the same time building a respectable library for the use of its members.[44] If by the 1890s the Mitchell Society was little more than a local university body, its founders, the vanguard of a new generation, had planted the seed of scientific organization in the state that would sprout in 1902 with the formation of the North Carolina Academy of Science, the first truly modern scientific organization in the South.

The Alabama Industrial and Scientific Society

Few southern scientists enjoyed the opportunities that the Mitchell Society afforded their Chapel Hill colleagues. Prior to the turn of

this

the twentieth century, only one other organization similar to the
Mitchell Society emerged in the South—the Alabama Industrial
and Scientific Society, organized in 1890 in Tuscaloosa. If Francis
Preston Venable served as the driving force behind the North Carolina scientists' early bid for professional organization, Eugene
Allen Smith filled that role in Alabama. Affectionately known
throughout Tuscaloosa as "Little Doc," because of his small stature, Smith enjoyed teaching and commanded the respect and ad-
miration of students and faculty alike. As soon as classes
adjourned for the summer, though, Smith, frequently in company
with his youngest son, Merrill, and a servant who doubled as
driver and cook, would set out in the "Little Doc's" ambulance, as
the state geological survey's specially designed Studebaker wagon
was known. Looking not unlike a medicine man or a peddler, and
once forced to explain at gunpoint that he was not a "revenuer,"
Smith spent thirty-one summers traversing the terrain of Alabama
examining mineral deposits and geological formations, mapping
various regions of the state, and collecting samples for the museum he organized. Smith made his last summer excursion in
1904. When progress blessed the survey with a Ford automobile
and the ambulance was retired, Smith left fieldwork to a younger
generation. Until his death in 1927, however, he continued to ana-
lyze the data provided by field assistants and to publish the results.[45]

Smith believed that the results of his work belonged to all the
people of the state. In addition to the numerous formal reports of
the survey, he published scores of articles in various newspapers
throughout Alabama that detailed his findings. He gave his advice
to all who sought it, including such budding Alabama industrialists of the 1870s as Truman H. Aldrich, Henry F. DeBardeleben,
and James W. Sloss. Although the mineral wealth of Alabama had
been tapped to some extent before the Civil War, many of the
fields soon to provide Birmingham furnaces with iron ore and
coking coal were, according to Ethel Armes, "untouched by the
pick of the miner when Dr. Smith directed attention to them."[46]
Throughout the 1870s and the 1880s the close relationship between
Smith and these entrepreneurs proved a boon to Alabama's economy.

industry

With the growing student body at the University of Alabama and the increasing demands of the geological survey, Smith requested an additional chemistry instructor. In 1890 the university employed William B. Phillips, Venable's first graduate student at the University of North Carolina and one of the founders of the Mitchell Society, as professor of chemistry and metallurgy. Upon the completion of his degree in 1883, Phillips had spent two years as a chemist for the Navassa Guano Company in Wilmington, North Carolina, and two years as professor of agricultural chemistry and mining at Chapel Hill, until the state shifted the land-grant support to North Carolina A & M College in Raleigh and Phillips lost his position. Phillips then migrated to Birmingham, Alabama, where he operated a consulting firm prior to his appointment at the university. No doubt Smith and Phillips discussed the Elisha Mitchell Scientific Society, for in addition to Phillips's association with the organization, Smith had been a corresponding member since 1885. Deciding that Alabama scientists could profit from a similar organization, the two colleagues presented their idea to the rest of the University of Alabama faculty, who not only approved it but joined them in calling for an organizational meeting, to be held December 11, 1890, in Tuscaloosa.[47]

Despite its inauguration by academics, the Alabama Industrial and Scientific Society (AISS) differed markedly from the Mitchell Society. Smith's close relationship with the iron and coal magnates led to their inclusion in the organization, which from the beginning espoused as one of its primary purposes "discussing various questions of interest to the material progress of the State." Response to the six hundred circulars distributed throughout Alabama included twenty-nine persons who attended the December meeting and an additional twenty-nine who expressed an interest in joining such an organization. Of these fifty-eight individuals, only ten of them indicated affiliation with an institution of higher learning.[48]

The inaugural address of President Cornelius Cadle, general manager of Cahaba Coal Mining Company, illustrates the conceptual difference between the AISS and the EMSS. While Venable had stressed the need to stimulate scientific research "for truth's sake" and the importance of encouraging young minds to consider a scientific profession, Cadle emphasized other values. "The reason for the creation of this Society," he maintained, "is found in the extraordinary development of the mineral region of our State, and in the fact that the association of scientific and practical men, for mutual help and interchange of experience, has always proved of great material advantage." He compared the AISS not to such organizations as the AAAS or the EMSS but to national societies of engineers.[49]

To the forty persons present at the first formal meeting on January 28, 1891, Cadle proclaimed that "science is the handmaid of economy," and he averred that "the time has passed when the 'Rule of Thumb' can be profitably applied to industry. Competition has made necessary the application of improved methods in all branches of production." Acknowledging the significance of the work of the geologist, the chemist, and various engineers in the advancement of iron production, Cadle concluded, "To the end that success may better attend our efforts, the men of science and the men of industry are here today."[50]

The AISS did not meet nearly as often as the Mitchell Society, usually convening semiannually in May and November throughout the 1890s. Although attendance at the first few meetings encouraged the organizers of the society, participation soon dropped to less than twenty regular members. In some instances a quorum necessary to transact business could not be mustered. Programs proved difficult to plan, for the majority of the membership— owners and managers of mining and manufacturing concerns— had little time to prepare even sketchy notes on their experiences with various processes and pieces of equipment. Frequently, persons scheduled to present a paper insisted at the last minute that they needed more time to complete their work. Others backed down altogether. H. F. Wilson, who had agreed to prepare a paper on the Birmingham Water Works, was "obliged to excuse himself because of the impossibility of doing this work in addition to what he has had to do in the office."[51]

Nor was the AISS successful in attracting new members. Because of the organization's narrow focus, its base of support centered in the industrial Birmingham region, and most persons interested in the society had joined at its inception. The AISS never expanded its horizons to include other branches of scientific investigation, and surprisingly, Smith and Phillips exerted no great effort to recruit their academic colleagues; thus a potential, if small, membership pool throughout the state remained untapped. Undoubtedly transportation difficulties also played a role in the geographical limitations of the AISS, much as they did for the EMSS.

Even though the meetings frequently proved to be a disappointment, a faithful core of members continued the organization and struggled to publish the *Proceedings of the Alabama Industrial and Scientific Society*, which appeared irregularly either once or twice a year. Each issue contained minutes of the infrequent meetings as well as several formal papers and occasionally the annual presidential address. The published papers, like the programs, reflected the interests of the coal and iron industry, although occasionally a university scientist presented a paper and published it in the journal. Even then it was usually the work of either Phillips or Smith, whose topics reflected the dominant interests of the group: "The Ultimate Composition of Some Alabama Coals"; "The Clays of Alabama"; and "Improvement of the Iron Ores of This District."

Although the membership roster for 1899 indicated fifty-one active persons, Smith, the secretary, sounded a somber note when he reported that the meetings of the previous year lacked much in the way of formal papers. Not to be discouraged, however, he insisted that "we have had discussions that have been of interest and profit, and the promise for the future seems to be better than it has been for many years."[52] Smith's remarks reflected wishful thinking rather than reality, for the AISS convened for the last time on March 14, 1900.

While the conversation relative to the continuation of the society at this final session was not recorded in the official minutes, a reporter from the *Birmingham News* did attend and pinpointed what the members saw as the particular problems with the organization. Some felt that the meetings should be held in the evening and opened to the general public in order to generate more interest in the society. Others, including Truman Aldrich of the Cahaba Coal Mining Company, noted that the industrialists who constituted the bulk of the membership simply did not have enough time to devote to the organization. Still others argued that teachers in the colleges and secondary schools with an interest in scientific matters should be induced to join and that the society should invite well-known scientists to lecture on topics of general interest.[53] All suggestions for survival came to naught, and the AISS never reconvened.

Both the EMSS and the AISS represented transitional institutions for southern scientists who sought national as well as regional professional recognition. They realized that to gain such status they not only had to engage in scientific research, but they had to attract the attention of their peers as well. This meant presenting papers at professional meetings and publishing research results in respectable journals. Certainly the Chapel Hill organization was attuned to these specific needs and attempted to meet them at least in part with the contact and publication outlet that it provided. However, the Chapel Hill scientists recognized the provincial nature of their society and understood that it alone could not catapult them into the national mainstream.

The AISS, although initiated by two scientists, stressed different goals, more akin to the immediate needs of the coal and iron interests of north Alabama. While this organization, too, met one of the needs of professional scientists, that of providing service to the community, this was not as important at the end of the nineteenth century as were personal contacts and publications. Consequently the AISS attracted few members from the academic community, and the industrialists who joined could not afford the time necessary to keep the society alive. The depression of the 1890s, from which Alabama's industry did not recover for decades, only aggravated the situation. In short, the organization was not crucial to them, and when it became more of a burden than a convenience, they discontinued their support.

Most southern scientists, of course, remained oblivious to either the EMSS or the AISS. Thinly scattered across the region, employed largely by colleges that could scarcely afford their salaries (let alone expenditures for equipment and travel), possessed of only a bachelor's or a master's degree, the vast majority of these persons had not been directly exposed to the emerging sense of professionalism. They utilized their education not in conducting original research but by passing their knowledge on to under-graduate students. Slowly, though, with the passing of each aca-demic year, southern scientists increased their contact with their counterparts from other regions of the nation through meetings, correspondence, and job mobility. By the turn of the century, those aspiring to true professional status desperately needed more than organizations such as the EMSS could offer. State academies of science, commonplace in other regions by this time, were just over the horizon in the South.

The Need for Organization: Emergence of State Academies of Science

3

In the years following the Civil War, many scientists in other areas of the nation experienced the same intellectual and geographic isolation as their southern colleagues. The rapidly expanding universities of the Midwest and West offered employment to trained scientists, but these young scholars, having migrated from more populous eastern cities, soon discovered the drawbacks of isolation so eloquently expressed by Francis Preston Venable. Anxious to further their careers, a handful of scientists in each of several states sought to overcome this predicament by organizing statewide academies of science. By 1900 eleven such organizations had been formed; twenty years later seven more had emerged; and by 1940, thirty-five state academies of science existed. Of these, eleven were in the states of the former Confederacy. Judging by their proliferation, these regional societies, despite being interdisciplinary in an age of increasing specialization, must have held much promise for the scientists who devoted long and sometimes frustrating hours to their creation.

Early Academies of Science

Several state academies of science claim to date from the late eighteenth and early nineteenth centuries. Usually they evolved from precursor organizations modeled along the lines of general learned societies such as the American Philosophical Society of Philadelphia and the American Academy of Arts and Sciences of Boston. In 1797, for instance, a few learned residents of Maryland organized the short-lived Academy Society; twenty-five years later the founders of the Maryland Academy of Science insisted that its heritage properly began with the Academy Society. However, this new Maryland academy languished when fire destroyed its building and collections in 1834. An effort two years later to revive the academy failed; not until 1867 was the current Maryland Academy of Sciences chartered.[1]

The New York Academy of Science dates its founding to 1817 and the organization of the New York Lyceum of Natural History. Although the word "lyceum" brings to mind evening lectures designed to educate the public, this society, created primarily by faculty of the College of Physicians and Surgeons, seems to have had as its purpose the benefit of the members. The lyceum met frequently in rooms provided in a city building and assembled a significant specimen collection and a less noteworthy library. In 1823 it inaugurated a series of technical publications which appeared irregularly as *Annals, Proceedings, Transactions,* and *Memoirs.* By 1876 major changes had occurred in the society: the specimen collection perished in a fire, and the base of membership had shifted to Columbia College. At this time the organization changed its name to the New York Academy of Sciences, having decided that "lyceum" and "natural history" "no longer indicated the actual scope of the society's activities." In 1891 the academy originated the Scientific Alliance of New York, a loose affiliation of a number of specialized societies centered in New York City. Sixteen years later most of these societies formally affiliated with the New York Academy, rendering it one of the largest and most active of the state academies in America.[2]

As Americans moved west during the nineteenth century, they carried cultural as well as personal baggage. Towns grew around settled areas; so, too, did organized activities, including local scientific societies. Cleveland (Ohio) Academy of Natural Science (1844), the State Natural History Society of Illinois (1858), the

Chicago (Illinois) Academy of Sciences (1857), the Peoria (Illinois) Scientific Association (1875), the Academy of Science of Southern Illinois (1876), and the Brookville (Indiana) Society of Natural History (1881) all attest to the popularity of these local organizations. Educational institutions were part of the westward migration, as well. Colleges brought to the area faculty who, cognizant of the trend of professionalization, realized the need for less fragmented organization and created statewide academies of science. These societies usually met twice yearly, once for a field trip and once for formal presentation of papers. They published their proceedings and frequently complete papers as well, although financial limitations restricted the size of the journals. In short, midwestern scientists sought to create for themselves the benefits of professional organization.[3]

Creating a viable state academy of science was not as easy as it might at first seem. The emergence and growth of the Kansas Academy of Science offers a good case study of the goals of the founders of such organizations and the pitfalls they encountered. J. D. Parker, the driving force behind the Kansas academy, was a native of Illinois who in 1867 had accepted an appointment as professor of natural science at the fledgling Lincoln College (now Washburn University) in Topeka, Kansas. Kansas was hardly an auspicious state in which to launch an academic career. Of course life there had been quite exciting during the turbulent 1850s, when no one was quite sure just exactly how many people did live in the territory and local residents battled one another as well as Missourians over whether or not slavery should be allowed by the state constitution. Although Kansas had not been a major battleground during the Civil War, when Parker arrived in Topeka he could still see rifle pits south of town. The area's first railroad had arrived just two years before, and safe travel through much of the state depended on the good will of the Indians.

Having been a member of the Illinois Natural History Society, Parker soon discovered that the absence of such an organization left a void in his professional life. He missed close contact with his colleagues and the organization's modest but current scientific library, for Lincoln College's collection was woefully inadequate.

He discussed his plight with B. F. Mudge of the State Agricultural College in Manhattan and broached the idea of a scientific society in Kansas. Mudge liked the suggestion but astutely pointed out to his compatriot that few scientists lived in the state; perhaps the time was not quite ripe. Not to be dissuaded, Parker eventually won over the skeptical agriculturalist.[4]

An organizational meeting for the Kansas Natural History Society convened in Topeka in September 1868; only a handful of people attended, no papers were presented, and Mudge was the only person not a resident of Topeka. The second meeting fared little better. However, three professors at Kansas State University invited the society to convene in Lawrence in 1870, and this meeting proved to be the turning point. The following year the membership voted to expand its scope by including "every line of scientific exploration and investigation" and to change the society's name to the Kansas Academy of Science. Then in 1873 a surprisingly generous state legislature not only authorized the publication by the state printer of the academy's proceedings but also donated space in the capitol building for the permanent quarters of the society. The Kansas legislature extended its support of the academy in 1895 by granting eight hundred dollars to the growing organization, an allocation which increased over the next twenty-five years.[5] Thereafter the Kansas Academy of Science attracted members primarily from colleges across the state who were anxious to expand their professional horizons.

Over the subsequent few decades scientists throughout the nation organized similar statewide academies; membership represented primarily the ranks of college professors rather than the small but growing body of industrial research scientists. Although laypersons were not usually barred from membership, few joined, for the papers presented at annual meetings and published in the journals encompassed detailed research results that were a far cry from the lyceum lectures of an earlier era. Some of the academies did acknowledge a continued responsibility to disseminate information to the public; occasionally they sponsored public lectures that addressed such popular subjects as Indian burial mounds and astronomical or weather phenomena. Generally, however, their primary concern was to serve as a focal point for member scientists—to offer them contact with one another through annual meetings and modest publications.

It would be erroneous to assume that all scientists statewide flocked to these academies; especially those with national reputations viewed these organizations as secondary to the AAAS and national speciality societies, although they usually joined and sometimes participated in meetings. Joseph LeConte noted of the California Academy of Science (founded in 1853) that it "had little effect in determining my scientific activity. I read many papers there, to be sure, and several of them were published in their *Proceedings*, but I always reserved the right to publish them elsewhere also."[6] While this situation was true of state academies throughout the nation, these organizations nonetheless enjoyed at least some support from prominent scientists and offered much-needed professional contact to lesser lights. It would remain to be seen whether or not the state academies could attract into membership nonacademic scientists as well, those employed by state governments and industrial research facilities.

The North Carolina Academy of Science — *Historical overview*

Among the southern states, North Carolina was the first to have a state academy that has remained in continuous existence. (Scientists in Texas organized a short-lived academy in 1892; it was not revived until the 1920s.) The origin of this academy offers a case study of the predicament of hundreds of scientists scattered throughout the nation. Their correspondence reveals frustration similar to that expressed earlier by J. D. Parker over their lack of contact with one another, a sense of isolation from the national scientific community, and the need to promote research and find a publication outlet. Some of them knew of the Elisha Mitchell Scientific Society, but in 1892 that organization had limited its membership to persons affiliated with the University of North Carolina in Chapel Hill. Consequently, scientists who taught at other colleges throughout the state and those who worked with the state agricultural extension service, the department of agriculture, and the United States Forest Service remained isolated from their colleagues.

Credit for first introducing the idea of a statewide academy of science goes to William Willard Ashe, who in 1901 suggested the undertaking to both Franklin Sherman, the state entomologist and later professor of zoology at Clemson College (later Clemson University), and Herbert Hutchinson Brimley, curator of the state museum.[7] Understandably, Ashe felt more isolated than most scientists in the state, for as forester for the North Carolina Geological Survey he was not affiliated with any of the colleges or universities. Nonetheless, he was an accomplished scientist. Born in Raleigh in 1872, Ashe attended the University of North Carolina and in 1892 received the master of science degree from Cornell University. In 1905, following his service to the state of North Carolina, he accepted a position with the United States Forest Service and continued in that post until his death in 1932. Throughout his career Ashe expressed great concern over the destructive effects of forest fires and lobbied for stronger state and federal regulations regarding the use of government-owned forests.[8] He advocated the wise use of forest resources by planting commercial stands of longleaf pine throughout North Carolina and introducing several improved extractive techniques to the turpentine industry.[9] Ashe retained membership in the North Carolina Academy of Science throughout his life. While seldom in attendance at the annual meetings after he left the state, he often published articles in the academy's journal.

Although Ashe originated the idea of an academy, Brimley's energy brought the organization into existence. Born in England in 1861, Brimley immigrated with his family to North Carolina in 1880, following a chance encounter with an immigrant agent who convinced the family that low farm prices in Europe should induce them to try their luck in America. During his first year in the United States, Brimley unsuccessfully worked a small farm and then taught briefly in a one-room school. Largely self-educated, Brimley turned his interest in the natural world into a business when he and his brother, C. S. Brimley, established a taxidermy service. For over twenty years Brimley Brothers, Collectors and

Preparers, supplied specimens for educational institutions and met the needs of private collectors. In 1894 the state museum of natural history, at the time a very small operation under the wing of the state department of agriculture, hired Brimley to reconstruct the skeleton of a fifty-foot right whale from an unorganized pile of bones. The next year he was appointed curator of the museum, a position that he retained until his retirement in 1937. Under his leadership the museum expanded from two thousand square feet to more than thirty thousand square feet. With his brother and T. Gilbert Pearson of the State Normal College (later the University of North Carolina at Greensboro), he coauthored *Birds of North Carolina*; he and Hugh M. Smith coauthored *Fishes of North Carolina*.[10]

In early March 1902 Brimley, Ashe, Sherman, and F. L. Stevens, professor of biology at North Carolina A & M College (later North Carolina State University) met together in Brimley's office to discuss the feasibility of forming a state academy of science.[11] The men agreed to contact their colleagues throughout the state in search of support for a proposed "state organization of those interested in scientific subjects." As Brimley wrote to William Louis Poteat, professor of biology at Wake Forest College, "It seems desirable from every point of view to form some sort of permanent organization which will bring the scientists of the State into closer contact with each other and stimulate research and the study of the local scientific features." He assured Poteat that although the organization had been conceptualized by a group of Raleigh biologists, "We are especially anxious to have a full representation of those outside of Raleigh."[12]

In a more personal letter to his good friend T. Gilbert Pearson, Brimley mentioned that biologists living in Raleigh had already formed a local organization "that is meeting with a good measure of success." This group, he continued, had decided to encourage "an organization of workers in all scientific lines broad enough to cover the whole state," and he called for a preliminary meeting at the state museum on Friday, March 21, 1902. "Now," quipped Brimley, "please take notice that unless death or marriage intervenes your presence is expected to help give the movement weight." Brimley then mentioned to Pearson that one of the hopes of the founders was that the new society could publish a journal "for recording the work of the members particularly in regard to local matters and if its object extended no further I believe that it would fill the usual long felt want." He closed by emphasizing once again that the proposed society was to be "a state affair and not merely an annex to the Raleigh Biological Club."[13]

On March 20, one day prior to the scheduled meeting, Ashe, Stevens, Sherman, and H. H. Brimley, along with Tait Butler, the state veterinarian, J. L. Kesler, professor of science at Baptist Female University (later Meredith College), and C. S. Brimley held an informal caucus in Sherman's office to determine the structure of the proposed organization. Opting for a state academy that would encompass all sciences, these biologists expressed a concern that chemists, by virtue of their numbers, could dominate the body. (Many chemists held positions with the state agricultural extension department, located in Raleigh, as well as with educational institutions.) Evidently the possibility that the Raleigh biologists might outnumber everyone else did not arouse similar anxiety. At any rate, the organizers decided to invite only Benjamin Wesley Kilgore, the state chemist, and W. A. Withers, professor of chemistry at North Carolina A & M, to convene with them on the following day.[14]

The seven men then delineated two classes of membership for the proposed organization. Fellows must be nominated by two other fellows and elected by a majority of the executive committee, primarily by virtue of their professional accomplishments. Regular members, elected by a majority of the executive committee, could participate in meetings but would not be permitted to vote or hold office.[15] Such a distinction seems elitist and so outraged a number of members that the provision was altered within three years. However, the founders of the academy intended that the primary focus of the organization be that of meeting the needs of professional scientists; while they were willing to admit to membership any interested individual, and in fact knew that the financial support of these persons would be essential, they feared that political participation by perceived nonprofessionals could dilute their goals. In their minds, the North Carolina Academy of Science was not to become a garden club, and its planned publication must not be a series of reminiscences of Saturday afternoon nature hikes.

With preliminary organizational matters in hand, the caucus concluded its work by drawing up a slate of officers: Henry Van-Peters Wilson, professor of biology at the University of North Carolina, was nominated for president; Pearson was chosen for vice-president; and Sherman was selected for secretary-treasurer. Nominated to the executive committee were Stevens, Butler, Kesler, Poteat, H. H. Brimley, Ashe, and either Withers or Kilgore (since neither of these chemists had been invited to attend the caucus, those present decided to see if either man expressed an interest). No one noticed—or mentioned—that all of the proposed officers, except for one person on the executive committee, were biologists. Finally, these seven men agreed that a primary consideration of the new society, in addition to regular meetings, should be the publication of a journal.[16]

On the following day nine persons assembled for the first formal meeting of the academy, six of whom had attended the caucus (Butler was out of town). In addition Poteat, Pearson, and Kilgore arrived to lend their support. Hubert A. Royster, a Raleigh physician soon to become head of the School of Medicine at the University of North Carolina, had agreed to join but was unable to attend the meeting. Although the caucus had chosen Wilson as its nominee for president, his absence led to the nomination from the floor of both Poteat and Stevens. After the first ballot, which resulted in a 4–4 tie between Poteat and Wilson (Stevens received one vote), Poteat was elected by a margin of 6–3. Pearson and Sherman were quietly elected. Kilgore then "made a resolute fight" to leave several positions on the executive council unfilled until future meetings of the academy. He lost the battle, and the nominees (including Kilgore) were elected.[17]

The constitution adopted at this meeting closely resembled that recommended by the caucus, although Kilgore insisted that five dollars was too high for dues, and he opposed the idea of a publication. An interesting technical question arose when the group discovered that, having adopted a constitution that allowed only fellows to vote, no fellows had been selected. Over Kilgore's vigorous protests, the executive committee hastily adjourned to elect fellows and chose the nine men in attendance plus Wilson, Royster, Butler, Withers, and J. I. Hamaker, professor of biology at Trinity College (later Duke University). Meanwhile, Kilgore left the meeting.[18]

Benjamin Kilgore's opposition to some of the organizational aspects of the society foreshadowed a chronic problem faced by many of the state academies of science—that of attracting the interest of scientists outside the academic world. Kilgore, although closely affiliated with North Carolina A & M College as a sometime chemistry professor between 1903 and 1907 and dean of the College of Agriculture from 1923 to 1925, saw his primary responsibility as service to the state's farmers. As state chemist from 1899 to 1919, director of the North Carolina Extension Service between 1914 and 1915, and director of the Agricultural Experiment Station between 1901 and 1907 and again from 1912 to 1925, Kilgore's work aimed at improving agricultural methods and educating farmers in their use. He was well respected for his pioneering research, which included studies on the relationship between cattle feed and milk production as well as the varied use of fertilizers.[19] Kilgore's career thus hinged on practical implementation of his research rather than publication, and he did not need a state academy of science, with its proposed journal and elitist membership, to further his career. Interestingly, he maintained his membership in the academy, although he seldom attended meetings, never served as an officer, and published nothing in the journal. Kilgore's position foretold that of industrial scientists who, in subsequent decades, failed to exhibit much interest in the state academies. However, the majority of the membership, college professors, desperately needed the contact with their colleagues and the publication outlet, and it was these men and women who shaped the future of the organizations.

On March 23, 1902, the *Raleigh News and Observer* carried an announcement of the organization of the North Carolina Academy of Science, maintaining that it "is likely to be far reaching in its effects on the development of scientific study and research within our borders." Probably written, or at least dictated, by one of the charter members, the article outlined the primary objectives of the academy to be the promotion of research in all scientific problems "and to encourage original investigation particularly in regard to native phenomena." Papers presented by members "deserving permanency among the literature" would be published in a journal which would be "representative of the best scientific thought and work done in the State." Intent on establishing their professional reputations and promoting the new society, the members declared that "no effort will be spared to make it [the journal] of the highest type of effort and reliability." The enthusiastic author of the article noted the "exceedingly satisfactory membership" and maintained that the academy "will very shortly embrace all the prominent workers in the various lines included in its plan and scope."[20]

After two postponements, the North Carolina Academy of Science convened for its first annual meeting on November 28–29, 1902, on the campus of Trinity College in Durham. At this two-day meeting, members read fourteen papers and presented ten others "by title only." The academy did not invite the public to attend, nor did it offer any lectures of general interest. Although the executive committee elected twenty-three new members, attendance was not as good as had been expected, and the organization's limited financial resources (dues were five dollars for fellows and one dollar for regular members) did not allow for the publication of a journal that year. The executive committee acknowledged that, without some changes, the academy would wither from lack of support.[21]

A solution to the financial stricture that blocked the publication of a journal soon appeared. The Mitchell Society, itself in financial straits, offered the pages of its journal to the North Carolina Academy of Science if the latter would bear two-thirds of the increase in publication costs. Depending upon how many articles the NCAS submitted, this arrangement would cost the academy about one hundred dollars annually. The membership voted to accept the offer, for while it preferred a publication under its own control, financial realities dictated that at present the Mitchell Society's proposal was the only means available to bring research results to print.[22] Eventually the two societies agreed to share editorial responsibilities as well as expenses, an arrangement that continued until 1983, when the EMSS formally ceased to exist and the academy assumed complete responsibility for the journal.

Another problem faced by the executive committee was the dissatisfaction expressed by potential members concerning the division between fellows and regular members. Hoping to enhance interest in the academy, the fellows in 1903 voted to amend the constitution and eliminate the category of fellow. Subsequently, any person interested in science and in the work of the academy could join upon nomination by two current members, a majority vote of the executive committee, and payment of annual three-dollar dues. Persons not desiring full membership status could become associate members upon the payment of one-dollar dues. Associates could attend the annual meetings and received a copy of the proceedings, but they could not vote or hold office.[23]

Even with these changes the membership grew slowly. By 1909 only ninety-four people had ever joined the academy, and many persons remained on the roll only a year or two (eleven of the initial membership had left the organization by 1904). Attendance at the annual meetings disappointed the faithful, even though the academy carefully avoided charges of localism by rotating the place of meeting among Raleigh, Chapel Hill, Durham, Greensboro, and Wake Forest.

In an effort to broaden its base of support, the academy invited local scientific societies such as the Audubon Society of Greensboro and the North Carolina section of the American Chemical Society to meet simultaneously with it and to establish a formal affiliation. By 1905 the academy also decided to offer an evening lecture and invite the general public. A newspaper article announcing the May meeting, probably written by the NCAS secretary Franklin Sherman, proclaimed that while the membership of the society was not large, "the attendance at meetings has been as good as could be expected, considering how widely scattered are the members, and the academy bids fair to continue a useful organization." However, continued the article, "it is hoped during the ensuing year to largely increase the membership."[24]

Such was not to be the case. Between 1906 and 1909, thirty-nine people joined the organization, but few of them remained active for any length of time. For example, twenty-one of these new members came in 1907; yet when the next full membership list appeared in 1913, only six of them were left. Although not explicitly stated, it is quite possible that during these early years some of the nominees for membership were unaware of their election to the academy until after the fact; if they chose not to participate, their name would appear on the roll for only one year. Attendance at the annual meetings was correspondingly low. A mere twenty-seven persons arrived for the 1909 Durham meeting; tellingly, twenty-seven papers appeared on the program. Fearful for the academy's future, the officers appointed a membership committee to spread word of the academy throughout North Carolina.[25]

The committee did its work well; when the academy met again in 1910, the secretary reported forty-five new members, for a total of eighty-nine on the roll. Not surprisingly, twenty-seven of the new members represented institutions of higher education: State Normal School (eight members), University of North Carolina (five), North Carolina A & M (five), Trinity (three), Wake Forest (two), and one each from Guilford, St. Mary's, Davidson, and Randolph-Macon (Va.). Fourteen members represented government agencies, primarily the North Carolina Board of Agriculture and the state agricultural extension service. In addition, one was a high school teacher, two were physicians, and one was the wife of a North Carolina A & M professor.

The new member from Randolph-Macon College was Ivey F. Lewis, a native of Raleigh who had received his A.B. and M.S. degrees from the University of North Carolina and his Ph.D. from the Johns Hopkins University in 1908. The grandnephew of Kemp Plummer Battle, president of the University of North Carolina from 1876 to 1891, Lewis retained an avid interest in educational progress in North Carolina and eagerly joined that state's academy of science in the absence of one in Virginia. Although his membership lapsed in 1912 when he became assistant professor of botany at the University of Wisconsin, he helped to found the Virginia Academy of Science a decade later, having returned to that state in 1915 as professor of biology and agriculture at the University of Virginia.[26] Two other members of the NCAS who eventually moved out of the state also worked to establish academies in their new homes. B. B. Higgins of the state agricultural experiment station joined the academy in 1910. Shortly thereafter he accepted a similar position in Georgia and was among the vanguard in founding the Georgia Academy of Science in 1923. Likewise, C. M. Farmer, a faculty member at Atlantic Christian College in Wilson who joined the NCAS in 1914, moved to Troy, Alabama, in 1920 and soon joined forces with other colleagues there to found the Alabama Academy of Science.

Hoping to increase participation at the annual meetings and especially anxious to retain the forty-five new members, the executive committee in 1910 altered somewhat the structure of the meetings by limiting presentations to fifteen minutes each and stipulating that no one could read a second paper until everyone who wished to do so had read a first one. Furthermore, a time for discussion following each paper was to be incorporated into every program.[27] All too often a handful of persons presenting lengthy and detailed reports had dominated the proceedings, discouraging many new members and boring supporters of the organization who had no interest in the topics on the agenda. Obviously, in an era when specialization was becoming the norm, regional, non-specialized societies faced a particularly ironic problem. On the one hand, they remained committed to the professional need of many members to present to their colleagues the results of their research. However, these reports were by their very nature intricate and detailed; even the best-educated scientists experienced difficulty following arguments in disciplines other than their own. To scientists with moderate training, many of the presentations were beyond their immediate comprehension, and the level of research thus exhibited discouraged them from offering comments on their own more modest efforts.

The immediate results of this new policy appear to have been positive. Although the membership level fluctuated, it never dropped below 68, and by 1920 was up to 112. Furthermore, people who joined the NCAS during its second decade tended to remain members for a greater length of time. Of the 107 people who joined between 1911 and 1920, only 15 left after just one year. The number of papers on the program climbed as well, from 22 in 1910 to 31 in 1911, before dropping slightly to 29 in 1912. Although this figure fluctuated along with the membership, declining significantly during the years of World War I, it rose to 33 in 1920 and skyrocketed soon thereafter.

Despite the primary interest of academy members in discussing the specifics of their disciplines, they did exhibit a concern for major issues of the time. A number of papers at each meeting addressed such topics as the public perception of the conflict between science and religion, the quality of science instruction at both the secondary and collegiate levels, and the conservation of natural resources. Scientists at all levels of professional development as well as the general public could relate to these issues, and they could work in harmony to correct misconceptions, to improve scientific education, and to encourage conservation principles.

In 1904 retiring NCAS president Charles Baskerville, professor of chemistry at the University of North Carolina, voiced two of these concerns to his colleagues. He first expressed dismay over the public perception that science and religion were natural antagonists, claiming that "the modern spirit of science towards religion is sane and healthy."[28] Undoubtedly he was referring to the controversy that had raged since the publication of Charles Darwin's *On the Origin of Species*. By the early twentieth century the vast majority of scientists throughout the world, many of whom were practicing Christians, had accepted the theory of evolution and experienced no intellectual conflict. These scientists were thus disturbed by the persistence of the general belief that the theory of evolution and the Christian religion, particularly the biblical account of creation, were mutually exclusive. At the time, Baskerville did not recommend any specific academy action, other than urging his colleagues to work to dispel what they viewed as a misconception. Little did he know that, twenty-two years later, a member of his audience, William Louis Poteat, would become the center of a violent confrontation between these two forces as fundamentalists sought to have him removed from the presidency of Wake Forest College for teaching evolution in his biology classes.

Baskerville also bemoaned the poor quality of science education in the state's educational institutions, at both the secondary and collegiate levels. He noted that while the number of science courses had increased significantly, the quality of instruction had not always kept pace with the rapidly expanding body of knowledge. Keenly aware that degrees in scientific subjects from southern colleges and universities were not held in high repute nationally, he pointed out that North Carolina graduates often could not find employment outside the state. The key to improvement, of course, was financial. Noting that the quality of instruction could be directly related to the intellectual excitement of the instructor, Baskerville maintained that colleges and universities must have adequate laboratory equipment and faculty research time. He also insisted that state governments, "far richer than ever in their history," should be willing to aid the very means by which they gained their newfound wealth. After all, he said, such riches followed "as a result of the progress of industries. Science sowed the seed of the present prosperity and it is worthy of remembrance, thanks, reward."[29]

Baskerville's address made no demands on the academy for immediate action. Rather, he seemed to have meant it as a pep talk, for he concluded by pleading with his audience "for greater activity in research." Admitting the importance of scientific work with a "practical" outcome, he nonetheless insisted that "pure science be either kept ahead or abreast of commercial progress" by perseverance in the academic laboratory. No one, he emphasized, could be a thorough professional in his or her chosen field without wishing to advance it. "Yes," he concluded, "our equipment is meager; poorer than it ought to be." But scientists should labor diligently with the resources at their disposal rather than wringing their hands in despair.[30]

While Baskerville's intention was to attract attention rather than incite specific action, other, more narrowly focused addresses to the academy membership did produce concrete results. In 1909 C. W. Edwards, professor of physics at Trinity College, read a paper entitled "College Entrance Requirements in Science in North Carolina" that painted a distressing picture of the inadequate academic preparation of the state's high school students. At the same meeting Mrs. Charles Duncan McIver, field secretary for the Woman's Betterment Association of North Carolina and widow of the founder of the North Carolina Normal and Industrial College (later Women's College, then the University of North Carolina at Greensboro), spoke on the efforts of that organization to improve North Carolina schools. These remarks "provoked considerable discussion" among the twenty-seven people in attendance and led to the formation of a five-member committee to collect more specific data concerning secondary instruction in the state and report the following year.[31]

As is often the case with committees of already overworked faculty, this one proved ineffective. The following year, it simply reported that it had not completed its task. However, the members of the NCAS did not let the matter drop. The 1910 meeting attracted the largest attendance to date, thanks to the membership drive, which netted forty-five new persons. Those assembled heard William Chambers Coker, professor of botany at the University of North Carolina, discuss "Science Teaching in the Schools and Colleges of North Carolina." Disturbed by his conclusion that much of the problem lay with poorly prepared teachers, the membership organized another committee "to collect data and plan courses of study in the sciences for the high schools of the State, to be submitted for approval by the Academy at the next annual meeting." Eight persons in addition to Coker accepted this responsibility, including representatives from Trinity College, St. Mary's College, Kinston High School, North Carolina A & M, and the State Normal College.[32]

In 1911 Coker presented the report of the education committee to the NCAS and subsequently published it in the *North Carolina High School Bulletin*. The recommendations included increased compensation and improved training for the state's high school science teachers. Higher education, asserted the report, must concentrate on preparing science *teachers* as well as science workers. Additionally, the committee suggested that the state board of education employ a scientist to plan the expansion of the science curriculum and that individual teachers not be expected to cover all scientific subjects.[33]

Interestingly, the committee concluded that science education in both the primary and secondary schools should "be made thoroughly practical, and that the study of the lessons of the textbooks be subordinated to the study of the phenomena themselves." Coker had emphasized the same point a year earlier in an article entitled "Science Teaching," noting that the great majority of high school students then did not attend college and that too often the theoretical aspects of education were soon forgotten. In fact, he maintained that for these students the curriculum should exclude mathematics beyond simple arithmetic, the history of ancient wars, and foreign languages. While he did not advocate the complete elimination of a liberal education, he did state that rural schools should concentrate on farming, gardening, and homemaking. Urban schools should follow the same principle, although the specific content would be somewhat altered.[34]

The membership of the NCAS did not share universally Coker's views on high school science education. Others, such as C. W. Edwards, maintained that the preparation of students who would apply for admission to the state's institutions of higher learning was of greater priority. Regardless of their orientation, however, many of the state's scientists agreed that North Carolina's young people needed a better foundation in the physical and biological sciences. Within a very few years the NCAS opened lines of communication with the state board of education and initiated a number of programs, including student essay competitions and open forums for high school teachers, that were designed to improve the high school science curriculum and to motivate students to pursue scientific studies (see chapter 6).

Another early interest of the North Carolina Academy of Science was conservation of natural resources. In 1907 Collier Cobb, professor of geology and mineralogy at the University of North Carolina, delivered the presidential address, entitled "The Garden,

Field, and Forest of the Nation." He pointed to the woeful conditions under which the state's farmers labored, including the credit system and exhausted soil, and noted that politicians biennially flattered the farmer to win his vote and then neglected agrarian interests in the legislative halls. Cobb maintained that scientists should respond to the needs of the farmer by bringing to his aid the "sensible science of our day, which has for its ultimate aim not merely discovery, but application, which is not so delighted with the formulating of a new law as it is overjoyed at the lifting of a burden." Through laboratory and field experimentation, scientists could expand humankind's knowledge of soil fertility and plant breeding, thereby reducing the farmers' dependency on change that all too often left him "poor in purse and lean in hope."[35]

While Cobb's concern for conservation addressed a specific need of the region's economy, Joseph Hyde Pratt, head of the North Carolina Geological Survey, advocated a somewhat broader program. While agreeing upon the need for soil improvement, Pratt stressed the issues of reforestation, the use of waterpower for industrial development, the reclamation of swamp lands, and the protection of ocean resources. According to Pratt, "the wealth of the nation could be nearly doubled if careful consideration were given to the utilization of these waste products and the prevention of waste of the products that are utilized."[36]

The strong utilitarian bent of these two men, in contrast to the emphasis of Charles Baskerville, is readily apparent; given the poor state of the southern economy, this concern is hardly surprising. Southern scientists knew full well that their professional dreams of funding and adequate time for experimental research were directly related to the economic development of the region. Thus they not only promoted the need to care for the region's renewable natural resources, but they also supported the right of industry to use judiciously such resources as forests, water, and minerals. The result, they hoped, would be a strengthened economy that could then support improved educational facilities and research laboratories.

Despite these concerns, the NCAS during its early years was unable to take much positive action on any issue outside the administration of the academy. The membership continued to grow over the next decade, though, and by 1923, when scientists in most southern states were just beginning to organize academies of science, the NCAS could boast a roster of 201 members, 63 of whom also belonged to the national American Association for the Advancement of Science. The NCAS had become a respected organization of professional scientists that during the 1920s and the 1930s would seek to meet the needs of its members to come together, to publish the results of their research, and to work collectively on such issues as education and conservation.

The Tennessee Academy of Science

For ten years the North Carolina Academy of Science was the only state academy in the South. Then in 1912 Tennessee scientists organized a similar society. George H. Ashley, the state geologist, suggested the idea to Charles Henry Gordon, professor of geology at the University of Tennessee, who then became the instigating force. Gordon, like his colleagues in North Carolina, was well aware of the evolving criteria of disciplinary professionalism. Having received a B.S. degree from Albion College in 1886, Gordon taught high school and then served as a principal in Keokuk, Iowa. Following a brief tenure as an instructor in the Academy of Northwestern University, he returned to graduate school and received the Ph.D. in geology in 1895 from the University of Chicago. Until 1903 he continued to work in the public schools of Iowa and Nebraska, taking some time to study at Heidelberg University in Germany. Following brief stints with the University of Washington at Seattle, the New Mexico School of Mines, and the United States Geological Survey, Gordon settled in Knoxville, Tennessee, as professor of geology and mineralogy. Throughout his career he spent many hours in fieldwork and was especially active with the Tennessee Geological Survey. Gordon retired from his work in 1931 and died in 1934.[37]

As a member of a number of national professional organizations, including the AAAS and the Geological Society of America, Gordon realized the importance to career advancement of regular collegial association and publication opportunities provided by society journals. Aware that most of his Tennessee colleagues did not have their institution's encouragement or financial support to further their professional development, he hoped to offset at least some of these impediments with a statewide organization of scientists. In a letter signed by eighteen scientists representing the University of Tennessee, Vanderbilt University, and the State Geological Survey, Gordon (the primary author) echoed sentiments expressed a decade earlier by the founders of the North Carolina Academy of Science. "There is," he wrote, "urgent need of a closer association of those interested in the study of the sciences and related branches in the State of Tennessee, and the time is ripe for an organization that will promote these interests." The objective of the society would be "the interchange of ideas and the coordination of the various scientific interests of Tennessee." Pointing to successful academies in other states, Gordon insisted that an annual opportunity to read and discuss papers would enhance the professionalism of Tennessee scientists. He concluded by calling for an organizational meeting on March 9, 1912, in Nashville.[38]

Gordon was encouraged by the initial response to this call. Those who gathered in the state capitol building on that Saturday morning included a few high school teachers, representatives of the state geological survey, and faculty members from Vanderbilt University, University of the South, Southwestern Presbyterian University, Cumberland University, and the University of Tennessee. At the morning session a committee was appointed to write a constitution; that afternoon the assembly adopted the constitution, elected provisional officers, and called for the first formal meeting of the academy to convene on April 6, 1912.[39]

The constitution as adopted in 1912 reflected the aspirations of the founders of the TAS, which in many respects resembled those of their counterparts in other state academies of science. Their stated objectives included the advancement of scientific research, the diffusion of knowledge, the promotion of communication among the state's scientists, the publication of research reports, and the development of "the material, educational and other resources and riches of the State." The constitution specifically provided for an official publication, the *Transactions of the Tennessee Academy of Science*, to "be published as occasion demands."[40]

Academy secretary Wilbur A. Nelson of the State Geological Survey did not record the number of people who attended the April meeting, held in the Carnegie Library in Nashville. Following the election of permanent officers, with Gordon chosen as president, Asa A. Schaeffer of the University of Tennessee explained the format for the proposed journal, and the membership voted hereafter to hold annual meetings on the Friday after Thanksgiving Day. In all, eleven people presented papers during the daylong session. The enthusiasm of those present must have followed them home, for by the end of the year Nelson reported that seventy-four persons had joined the organization.[41]

The Tennessee academy convened again in 1912, on November 29–30 on the campus of the University of Tennessee in Knoxville. The two-day program offered sixteen papers by fourteen persons, one of whom was a woman. Eight of the participants were University of Tennessee faculty members; one represented the University of the South; two were employed by government agencies; and the remaining three listed no formal affiliation.[42] Among the presentations was Gordon's presidential address, "Science and Progress in the South," which reflected sentiments expressed by Francis Preston Venable to the NCAS a decade earlier. While achievements "are a direct outgrowth of the labors of certain individuals," he noted, "contributory thereto were scores of lesser lights and influences without which success would have been impossible."[43] He went on to comment that organizations play a pivotal role in the personal interchange that is essential for such expansion of knowledge, challenging his audience to turn the TAS into a significant organization for the South.

Gordon offered several suggestions for the future work of the academy. "If it shall serve no other purpose than as a stimulus for scientific work among its members, it will amply justify its existence," he maintained. Annual meetings will offer the opportunity for "the interchange of views through papers and discussions," which, he hoped, would arouse enthusiasm and help "to overcome the inertia of dull routine and to ward off the insiduous [*sic*] approach of indifference to the weal or woe of the other man." However, the membership should also promote "every enterprise tending to conserve the material and human resources of our State." For Gordon such promotion meant support of the state's geological survey, concern for instruction in the public schools, promotion of the state's abundant natural resources, and the creation of a state museum.[44]

At that meeting the academy membership took action on two of Gordon's recommendations. It adopted a resolution urging the legislature to enact a law that would authorize the governor to create a conservation commission. This organization would oversee the use of the state's waterpower and forests to ensure that state resources benefited Tennessee and were not diverted "to the enrichment of other states." Moreover, the commission should "cooperate with the boards of trade and other civic bodies" to stimulate industrial growth.[45] Specific suggestions for touting the state's industrial promise included exhibits at such upcoming events as the National Conservation Exposition, scheduled for the following year in Knoxville; a national exposition for 1914 in San Diego; and the Panama-Pacific Exposition at San Francisco in 1915.[46]

Although flushed with idealistic enthusiasm, the academy had little muscle to flex. Resolutions are easy to pass, but in effect the academy took little positive action during its first decade beyond providing annual meetings that usually offered between ten and twenty papers each. In 1914 it did hold a joint meeting with the Middle Tennessee Educational Association. No doubt the papers, which included "Science in the High School," "The Claims and Place of Biology in Education," and "The Relations of the Sciences to the Teaching of Agriculture," interested those in attendance, but the minutes of this meeting do not indicate that any resolution was sent to the legislature, nor was a committee formed to pursue any specific matter.[47]

Neither was the academy immediately successful in its desire to publish a journal. Only two slender volumes of the *Transactions* appeared prior to 1926, one in 1914 and a second in 1917. The major obstacle, of course, was lack of funds. With a dues-paying membership that declined slightly to sixty-six persons in 1917 and no other source of funding, the academy could not meet the cost of an annual publication. In 1925 a membership drive quadrupled the number of names on the roster, and venturesome officers decided to try a quarterly publication. The first issue of the *Journal of the Tennessee Academy of Science* appeared in 1926; since that time the *Journal* has been published regularly, with most of its pages given over to scientific articles authored by the membership.[48]

Other Southern State Academies of Science

Not until the 1920s did scientists in other southern states follow the lead of their North Carolina and Tennessee colleagues and form state academies of science. There is no clear explanation as to why these two states of the upper South took action first. Earlier, smaller scientific societies had existed in both states (the Elisha Mitchell Society in North Carolina and the LeConte Scientific Society in Tennessee), but neither of these organizations directly stimulated the creation of the state academy. Furthermore, similar local societies existed in other southern states, as well, and for a brief time in the 1890s a handful of Texas scientists had organized themselves into an academy. Both North Carolina and Tennessee contained a number of colleges and universities, several of which offered a modest number of graduate degrees, and were perhaps more advanced in this regard than other southern states, with the exception of Texas and Virginia. Consequently, scientists with a stake in advancing their professionalism were most likely to be found in these regions. Finally, as is necessary with any project requiring considerable coordination, an interested handful of people—those most concerned with their professional careers—were willing to accept the responsibility for pushing these organizations into existence.

In 1920 Virginia biologists, led by Ivey F. Lewis, who earlier had briefly been a member of the NCAS, organized the Virginia Association of Biologists. Although the limited specialization kept membership figures low, the biologists were pleased with the idea of organization and in 1922 opted to expand into the multidisciplinary Virginia Academy of Science, soon to become the largest of the southern state academies of science. In the spring of 1923 Lewis collaborated with several of his colleagues on a letter to scientists throughout the state, encouraging them to attend the April meeting in Williamsburg and to take an active role in the society's transition. Although the number of people in attendance went unrecorded, eighteen persons presented scientific papers on this occasion. The organizers, hoping to encourage both those assembled and others who had indicated interest to sustain a state academy, invited William C. Coker, professor of botany at the University of North Carolina and former president of the North Carolina Academy of Science, to address the group. His remarks, "The Scope and Function of a State Academy of Science," assured the delegates that such an organization not only would benefit the research scholar but also would prove of inestimable value to those persons whose primary responsibility was teaching.[49]

Founded with 134 charter members, the Virginia academy expanded to 315 members by 1926 and to 500 by 1929. This large membership, in comparison with the other southern state academies of science, can be explained in part by the preponderance of educational institutions in the state. The University of Richmond, the University of Virginia, William and Mary College, the Medical College of Virginia, and Virginia Polytechnic Institute all contributed substantially to the academy's membership. Scientists located in the state's smaller colleges, such as Hollins and Randolph-Macon, participated as well. Furthermore, scientists in Virginia were geographically closer than their other southern colleagues to the mid-Atlantic and northern cities that most often hosted national professional meetings. Undoubtedly more Virginia scientists participated regularly in such national activities and transferred the spirit of professionalism to the state level. Why, then, they were initially slow to organize a state academy remains a matter for speculation.

The Georgia Academy of Science developed along strikingly different lines from those in North Carolina, Tennessee, and Virginia. On the surface, the initial call for an organizational meeting, issued on February 14, 1922, by members of the University (of

Virginia Academy of Science, 1925 (Courtesy of Virginia Academy of Science Archives, Virginia Polytechnic Institute and State University Libraries)

Georgia) Scientific Society, sounded familiar. There seems to be a need for an organization in Georgia that would bring the scientific men of the State together and that would promote research. It is thought that a single relatively strong organization would be better than a number of special societies.[50] But the letter proposed qualifications for membership that reflected the exclusivity of the National Academy of Sciences rather than the more democratic American Association for the Advancement of Science and other state academies. As R. P. Stephens, professor of mathematics at the University of Georgia, stated, membership in the Georgia academy was designed to serve as recognition of scientific achievement and would be restricted to those who had "made [a] noteworthy original contribution to science." Only scientists who had been "five years successfully engaged in some line of recognized scientific work" or "five years continuously a member of the science faculty of a college of recognized standing" would be eligible. Adding to the exclusivity of the organization was a membership ceiling of fifty persons.[51]

Not everyone was pleased with the proposed structure. Among those who objected was B. B. Higgins, formerly a member of the NCAS then employed by the Georgia State Agricultural Experiment Station at Griffin. Higgins, reflecting sentiments not unlike those Benjamin W. Kilgore had expressed twenty years earlier concerning the North Carolina academy, argued that the proposed membership strictures would defeat the very purpose for which the academy was being formed. "My idea," he wrote to a colleague, "was to make the organization as all-inclusive as possible, so as to increase the interest of the educated public in science and to impress them with the possibilities of scientific research."[52] The matter produced considerable discussion at the initial meeting, but the final constitution contained provisions almost identical to those outlined in the circular letter. Moreover, the membership was divided into twelve specialty groups. An applicant for membership who met all other qualifications could be admitted only when an opening occurred in his or her field.

For obvious reasons, the Georgia academy grew more slowly than the other southern state academies of science. To compound the problem even further, the executive council voted in 1928 that applicants for membership must first present a paper to the academy and that members who failed to present a paper at least once in a three-year span would be removed from the active roster. Nonetheless, scientists throughout Georgia sought to become members of the organization, and the academy slowly altered the initial membership strictures. In 1929 the GAS voted to allow for sixty members instead of fifty, and similar increases occurred periodically during the next two decades. Additionally, the GAS voted in 1936 to institute two categories of membership, fellows and members. While any interested resident of Georgia could become a member, only the fellows chosen according to the initial elitist criteria could vote or hold office. Not until 1949 did the academy remove all such restrictions and provide for open membership, after which time the organization grew rapidly.[53]

Although organized on a somewhat different principle than other state academies of science, the GAS exhibited interests similar to them. In the presidential address of 1924, W. S. Nelms, professor of physics at Emory University, proclaimed the scientists' concern for research:

Is the work of research and investigation in science appreciated and supported adequately in Georgia at the present time? Is there a single institution, educational or commercial, in this state that offers the facilities or encouragement for the proper development of original investigations? These questions must cause a smile or a tear on the face of those who have the least interest in this matter.

In a prescient remark indicating his belief that the state government should fund research, he stated, "What would the state consider a fair price for a recipe for the extermination of the boll weevil? A million dollars invested in biological researches right here where the weevil is a menace, might bring forth this much desired formula."[54] Nelms's price of $1 million would have been a bargain, considering the destruction wrought by the pest as it munched its way across the state later in the decade.

Alabama scientists, despite their geographic location in the deep South amid relatively few strong institutions of higher education, organized a state academy shortly after their colleagues in Virginia and Georgia. In 1923 Wright A. Gardner, professor of botany and plant pathology at Alabama Polytechnic Institute (later Auburn University), approached a number of scientists in the state about the need for an organization "through which ideas could be exchanged and certain results be given to the public." As a graduate student at the Universities of Michigan and Chicago, he had experienced the intellectual stimulation that results from collegial contact and hungered for such in his adopted state. He pointed to the many state academies of science then in existence and noted that they "have proven very satisfactory in other states and should prove equally satisfactory and beneficial in Alabama."[55] Gardner received a number of encouraging replies to his letter. An aging Eugene Allen Smith, however, remembered the fate of the earlier Alabama Industrial and Scientific Society and was "not at all sanguine about the success of such an undertaking."[56]

Despite his misgivings, Smith joined the organization and participated in the inaugural meeting by recounting the story of the AISS. Sixteen other persons also presented papers to the fifty-six scientists who gathered on April 4, 1924, in Montgomery. Initially the academy met as the science section of the Alabama Education Association, in part to increase attendance at the meeting but also because a number of persons interested in the organization were high school teachers who had inquired of Gardner about the possibility of presenting nontechnical papers. Earl E. Sechriest of Ensley High School noted that one of his "pet projects" was visual stimulation as a means of education and submitted a paper entitled "The Eye as a Window of the Mind." Gardner replied that "it does not seem necessary that all members of our organization present ultra scientific records, though results on investigations are desirable."[57] Although two years later the Alabama Academy of Science had separated from the Alabama Education Association and was primarily an organization of college and university faculty members, high school teachers continued to support it. Their concerns led the academy during the 1930s to take positive action to improve secondary science instruction, while academies elsewhere did little more than discuss the issue.

Both the South Carolina and Louisiana academies experienced

more modest beginnings. Organized in 1924 and 1927 respectively, their organizational meetings reiterated the need for scientists to meet together and present the fruits of their research. Although both bodies aspired to publish a journal of scientific merit, financial stringencies dictated that, prior to 1940, they would have to confine their activities to sponsoring an annual meeting and searching for a broader base of support.[58]

The Arkansas Academy of Science initially convened in 1917; eleven persons drew up an elaborate constitution, including five different classes of membership and seven standing committees. However, only nine more people joined the organization that year, and it did not meet again until 1933. Absences from the state because of World War I and "other duties" were blamed for the organization's poor beginning. In 1932 faculty members at the University of Arkansas served as the core for the reorganized academy, which took as its name the Arkansas Academy of Science, Arts, and Letters in an effort to attract a large membership. The academy grew slowly, though, attracting primarily academic scientists. In 1941 the membership voted to reflect this fact, dropping "Arts and Letters" from the official name of the organization. They also published the first volume of their journal, although subsequent issues would await the end of World War II.[59]

The Florida Academy of Sciences was founded in 1936, primarily by faculty members at the University of Florida. In May, about 100 people attended the initial meeting of the FAS, where twenty-one papers were presented. Another meeting in November attracted 125 people, with twenty-seven papers. By 1940 membership had reached 383 people; more important, the academy began immediately to publish an annual journal that in 1945 became a quarterly publication. On the occasion of the twenty-fifth anniversary of the FAS, longtime member E. Ruffin Jones, Jr., reflected that in the early days, most members were academic and research scientists. "Even in those days," he commented, "I don't believe that many people considered participation in the Academy an adequate substitute for attendance at national meetings. But Florida was a long way from the large metropolitan centers where most of the national meetings were held, there wasn't much money and we had to do the best we could."[60]

The Mississippi Academy of Science experienced a long birth process. Chartered in 1930, it did not officially convene for the first time until 1937. One issue of a journal appeared in 1939, but then World War II disrupted both the publication and the annual meetings. Conventions resumed in 1946, but the journal proved much more difficult to revive. Only six annual issues appeared irregularly between 1947 and 1960; since then the journal has been published on an annual basis.[61]

As is evident from the effort that southern scientists exerted to organize state academies of science, these men and few women felt the need of professional contact with their peers. Interestingly, as the academies experienced growing pains and as they searched for their niche in the professional world, one of their earliest predicaments was finding a broad base of support. While southern institutions of higher education were growing, they still did not attract overly large or exceptional faculties. Many instructors held only bachelor's or master's degrees and did not share the sense of professional urgency with their more highly educated colleagues. In many cases the few people who represented the organizing heart of the academies came to dominate them, a situation that many would-be members no doubt found discouraging. Finally, the few scientists employed by the South's fledgling industries, with a different set of criteria for career advancement, seldom supported the academies.

The problem thus confronting state academies in their earliest years was how to structure the organizations to meet the needs of the academic scientists who founded them and still attract enough support—financially and otherwise—to operate. Even at this early date, some scientists questioned whether these general organizations were not anachronisms in an increasingly specialized world. Those who had become active in the academies, though, refused to let them die of neglect, seeking instead a magic formula that would render them essential organizations to enough people to ensure their survival.

4 | ## Support for Fledgling Academies: North Carolina Scientists, a Case Study

Scientists throughout the nation, including those in the South, sought organization among themselves on a more intimate level than that provided by the national professional societies. By the 1920s, annual conventions of the latter had grown large and impersonal. Moreover, while the number of scientists had expanded exponentially, publication space in journals had not, leaving many competent people without adequate resources for career advancement. Scientists in the southern states faced other limitations, as well. The relative prosperity of the 1920s did not extend to all parts of Dixie, large portions of which remained at the mercy of a declining cotton culture, sharecropping, and low-wage, labor-intensive industry. The Great Depression of the 1930s only exacerbated the situation. Many colleges struggled just to remain open, and even state universities with progressive presidents faced limited budgets that could not be stretched to cover release time for faculty research and travel expenditures.

State academies of science were thus born of professional needs. Men and women imbued with the spirit of collegial contact, original research, and publication found themselves teaching in areas of the nation that were far distant from centers of professional activity and in colleges that could not underwrite the expense of their professionalism. Even as these regional organizations were taking shape in the early years of the twentieth century, though, much discussion centered on the usefulness of statewide, multi-disciplinary organizations in an era of increasing specialization.

The Need for State Academies of Science

By the 1920s, when most of the southern state academies of science were organizing, specialization had almost completely overtaken the sciences. The proliferation of national societies devoted to specific disciplines caused some scientists to doubt the validity of the more general AAAS and the regional, interdisciplinary organizations. By 1918 the enthusiasm that had produced the midwestern state academies of science had waned to such an extent that David D. Whitney, president of the Nebraska Academy of Science, noted that the members of his own academy showed "lively interest at the annual meeting, but [were] apathetic the remainder of the year." He asserted that "only a small percentage of the scientific people of the country are members of the various state academies," primarily because "the day has gone by when men interested in widely different special lines of research or activity can profitably meet for the common discussion of their interests."[1]

In an attempt to prove his point, Whitney circulated a questionnaire to the state academies of science across the nation and received a reply from twenty-five of them. Among those that responded, three reported their organizations as "dead"; four, in addition to Nebraska, indicated that the majority of their membership was "apathetic." The remainder, however, indicated either "lively" or "reviving" interest—hardly sufficient evidence to sound the death knell for state academies.[2]

Aware that the growth of national organizations would inevitably alter the regional societies, some scientists perceived this change as an opportunity rather than a disaster for the state academies. In 1915 T. C. Mendenhall of the Ohio Academy of Science chided colleagues who did not join the academy because national organizations "offer as good or better facilities for the accomplishment of the principal ends the academy has in view." He feared that scientists with their myriad of specialized societies would lose sight of the broad spectrum of human knowledge by "burrowing in a trench." While he did not advocate a return to the days of the "gentleman amateur," he did maintain that scientists should use some of their precious time to keep abreast of significant developments, whatever their field.[3]

Mendenhall offered several rationalizations for the continuation of state academies. First, he insisted, their annual meetings still represented an important source of contact and camaraderie for scientists. Local societies, maintained Mendenhall, were far better suited to that purpose than the AAAS, which had grown too large. Moreover, he added, nonprofessional scientists still played a role in the local organizations, and their interests should not be overlooked. Finally, he argued that academy members could advise state governments on legislation requiring scientific expertise, particularly in respect to the conservation of natural resources.[4]

Paul P. Boyd, president of the Kentucky Academy of Science in 1920, conducted his own survey among the state academies and concluded that the organizations did indeed have a role to play in the scientific community. Responding officials emphasized that many scientists who could not travel to national meetings seldom missed state gatherings. Furthermore, they declared, state meetings provided an ideal starting point for young scholars and served to counteract the growing myopia of the various subspecialties. The respondents also argued that state academy meetings offered a common ground for professional scientists and interested amateurs and could be mobilized for numerous services to the state.[5]

A further defense of state academies of science was raised by W. S. Bayley of the Illinois academy in his address to the Wisconsin Academy of Science on April 6, 1923. Bayley argued that state academies should not even try to become little national societies, for such would lead to "a duplication of work and consequently a waste of energy." Rather, he suggested that state academies could assist local planning agencies, the government, and private industry with whatever matters were of greatest local concern. They could also serve as educational agencies to the general public by emphasizing that "scientific conclusions are not biased and that they are never final, but are constantly being tested as to their correctness." Undoubtedly Bayley's remarks were precipitated by the Oklahoma legislature's recent passage of a bill forbidding public school teachers to instruct their students in the scientific theory of evolution. Similar action was pending in several other states, and Bayley, along with much of the rest of the scientific community, was genuinely alarmed at what he perceived as public ignorance of the scientific method.[6]

Finally, said Bayley, state academies should be concerned with the science curricula of the states' high schools and colleges. He illustrated his point by mentioning the recent efforts of the Illinois academy to improve scientific instruction. In order to prevent high school science teachers from becoming "penny-in-the-slot machines for retailing little scraps of knowledge recurrently at the ringing of an electric bell," Bayley declared, the Illinois academy encouraged them to attend the annual meetings and tailored certain sections especially to their needs and interests. The academy also solicited information concerning science clubs in the state's high schools and offered these clubs an opportunity to affiliate with the academy. He concluded by indicating that, with these changes, the coming annual meeting "will not be of the same high technical value as last year's meeting . . . , but a much larger number of persons will participate in it—and a much larger share of the papers will be presented by young persons."[7]

Obviously many scientists believed that the state academies could make a positive contribution both to their professional lives and to society. Southern scientists of the 1920s echoed these sentiments; most of the academies, only recently founded, had not yet encountered widespread apathy and enjoyed an enthusiastic, if sometimes small, leadership. Before delving further into the development of the state academies of science, it will prove useful to give close scrutiny not only to the men and women who constituted the southern scientific community of the 1920s but also to their participation in the state academies of science.

Who Were These Southern Scientists?

In the years preceding World War II, the majority of southern scientists continued to be college and university professors. However, southern colleges and universities lagged behind their counterparts in other regions of the nation in terms of size, endowment, and graduate programs. Nor did the region support much research-oriented industry. Thus few topflight scientists sought positions in the South. This is not to imply, though, that southern scientists were not creditable professionals; an increasing number of them held graduate degrees and had been taught that research and publication were crucial elements of a scientist's life. A handful enjoyed a national reputation for outstanding contributions to their fields.

What attracted these men and few women to the educational institutions of the South? Perhaps most important, they responded to an increased availability of positions as a number of colleges and universities experienced considerable growth, especially during the 1920s. An influx of young men who had postponed their college education to serve in the armed forces during World War I generated a considerable portion of the increased enrollment. In addition, as state governments placed a greater premium on public secondary education, high schools graduated more students and consequently funneled more of them into the region's colleges.

At the same time, many of these expanding institutions endeavored to enhance their national academic reputations. Farsighted professional educators such as David C. Barrow at the University of Georgia and Harry Woodburn Chase at the University of North Carolina utilized increased funding from philanthropic as well as legislative sources to oversee their institutions' initial steps along the path to becoming true universities with an expanded emphasis on research and community service. Endowments such as that of tobacco tycoon James B. Duke to Trinity College in 1925 further enhanced the potential of southern institutions.[8]

Faculty size at southern colleges and universities grew also, from a pre-1900 average of approximately a dozen persons per institution to between sixty and seventy-five persons by the 1920s. Certainly science departments shared in this growth. At the University of North Carolina, a department of six science professors in 1893 expanded to twenty-five by 1925. During the same time period the science faculty at North Carolina State College grew from four persons to twenty-four, and that at Trinity College blossomed from one science professor in 1893 to eleven by the time of its transformation into Duke University in 1925. Smaller colleges experienced less spectacular growth, usually enlarging their science faculties from one person in the 1890s to three or four by 1920.[9]

North Carolina: A Case Study

The growth in numbers and increasing professionalism of these people becomes apparent upon a detailed examination of their presence in one geographic area. In North Carolina, between 1875 and 1940, a total of 498 persons taught in the physical and biological sciences (excluding such specialties as textiles and poultry science) at eight diverse institutions of higher education: the University of North Carolina, North Carolina State College, Trinity College (Duke University), Wake Forest College, the Women's College (the University of North Carolina at Greensboro), Meredith College, Elon College, and Davidson College. Table 4.1 shows, not surprisingly, that these scientists were concentrated in the two major state universities, although Duke University (especially after the endowment of James B. Duke in 1925) and Women's College in Greensboro also employed a significant number of scientists. The total number of persons in table 4.1 adds up to 524, because 24 of these persons served at two different institutions during this time and 1 person taught at three different schools.[10]

Table 4.2 outlines the growth by decade of the science faculties at each of these colleges. Not unexpectedly, the most significant increase occurred shortly after World War I, in conjunction with the expansion of colleges throughout the region. Veterans of World War I and an increasing number of academically minded high school graduates swelled the classrooms, while aggressive administrators sought able faculty and philanthropic and state funding to pay them. Again, the growth is most noticeable in the state-supported schools, although the benefit to Duke University of the endowment that it received in 1925 is readily apparent.

Table 4.1
Science Faculty in North Carolina, 1875–1940

School	Total faculty	Ph.D. or D.Sc.	Master's degree
Univ. of North Carolina	127	67	32
North Carolina State Col.	123	43	35
Duke Univ. (Trinity Col.)	80	53	14
Women's Col.	71	17	34
Davidson Col.	46	12	11
Wake Forest Col.	42	12	14
Meredith Col.	23	5	13
Elon Col.	12	6	1
Total	524	215	154

These North Carolina scientists exhibited a high degree of specialization in one discipline, although, as table 4.3 shows, no single field totally dominated the others.

During the period of greatest overall growth prior to World War II, 1920–25, a total of 162 persons (135 men and 27 women) taught physical and biological science courses in one of these eight institutions. Of these, 55 held an earned doctorate, 7 had either a D.Sc. or LL.D. (usually honorary), 1 had an M.D., 59 held a master's degree, and 40 had only a bachelor's degree. Many of those without the Ph.D. eventually earned one, indicating a sense of professionalism beyond the desire merely to teach on the postsecondary level. (Prior to World War II, few schools required the Ph.D. for their faculties.) A close look at the backgrounds and careers of these people reveals some surprises about North Carolina colleges and their science faculties and offers evidence as to why these persons needed and supported a state academy of science. Although North Carolina was one of the more progressive southern states at this time, these conclusions can with some degree of caution be generalized across the region.

First, more than half of these educators migrated to the South from other regions of the nation. While 62 of them never received an entry in *American Men of Science* and thus left no traceable career

Table 4.2
Distribution of Science Faculty in North Carolina, 1895–1935

School	School year								
	1895–96	1900–01	1905–06	1910–11	1915–16	1920–21	1925–26	1930–31	1935–36
Univ. of NC	6	8	12	12	15	18	25	27	27
NC State Col.	3	7	10	10	15	19	24	25	26
Duke (Trinity)	2	3	4	5	4	7	11	27	32
Women's Col.	4	2	3	4	5	10	16	16	16
Davidson Col.	3	2	2	3	4	6	8	13	10
Wake Forest Col.	3	3	3	6	5	9	9	7	6
Meredith Col.	0	1	1	1	2	4	4	7	7
Elon Col.	0	0	0	0	3	3	3	3	4
Total	21	26	35	41	53	76	100	125	128

Table 4.3
North Carolina Science Faculty—Discipline Specialization

Discipline	Total	Listed as one of several fields	Listed as the only field
Chemistry	354	179	175
Biology	189	100	89
Physics	189	99	90
Zoology	109	56	53
Botany	86	45	41
Geology	84	44	40

[handwritten note:] Is midgette only using for her data? See n.11 p.214

path, the birth states of the remaining 100 reveal that only 37 of them were born in the states of the former Confederacy. The source of their undergraduate degrees provides further evidence of this in-migration. Since most college catalogues of the 1920s listed each faculty member's educational history, this information is available for all but 15 of these persons. Of these 162 men and women, 70, or 43 percent, received their bachelor's degrees from a school in one of the states of the former Confederacy. Fifty of these 70 had attended a college or university within North Carolina. While the South produced a larger percentage of these scientists than any other single region of the country, it is readily apparent that a sizable portion of them entered the state from other areas of the nation. A total of 27 received their undergraduate education in the north central region (Ohio, Michigan, Illinois, Minnesota, Wisconsin); 26 more came from the Northeast (Maine, New York, Massachusetts, Rhode Island, and New Jersey); 12 attended schools in the mid-Atlantic region (Maryland, Pennsylvania, Kentucky, West Virginia); 9 came from the Midwest (Kansas, Nebraska, Missouri); 1 arrived from Oregon; and 2 of them came from foreign nations. (See table 4.4.) Obviously, southern institutions of higher education could attract faculty members from other regions of the nation.

Of course it is possible that southern-born men and women could have left the region for their undergraduate education, thus rendering the above figures invalid. Such a pattern was not un-

[handwritten marginal notes:] But you should throw out the 15 from the 162 from the total. 15 represents 9.3% of the total

[handwritten marginal note:] Of course i probably quite likely

Table 4.4
Regional Source of Undergraduate Degree, North Carolina
Science Faculty, 1920–1925

Region of nation	No.	%
South	70	43.2
North Central	27	16.7
Northeast	26	16.0
Mid-Atlantic	12	7.4
Midwest	9	5.6
Northwest	1	0.6
Foreign	2	1.2
Unknown	15	9.3
Total	162	

(handwritten margin notes: 147 in sample should be 47.6 / 18.4 / 17.7 / 8.2 / 6.1 / .7 / 1.4)

(handwritten: Unknown circled; Total circled)

common in the nineteenth century, at least before the Civil War, and historians have often assumed that it continued well into the twentieth century. However, the trend for these scientists tells a different story. Of the thirty-seven men and women known to have been born in the South, only one of them earned the bachelor's degree elsewhere. Thus, the fact that so many of these scientists received their undergraduate degrees outside the South is a reliable indicator that most of them migrated into the region.

Why so many people came South to teach, at least for a portion of their careers, is a matter of speculation more than of fact, for few of them recorded their reasons. For those without a doctoral degree, southern positions may have been more readily available than in many of the schools of the North and Midwest. A number of people earned a graduate degree at one of the North Carolina universities while teaching at the same time, no doubt an attractive option. Others probably saw teaching positions in one of the southern schools as a good starting point, a place to gain a few years' experience and enhance one's résumé before seeking another position elsewhere. Still others may have welcomed the challenge of participating in the growth process at schools such as the University of North Carolina, North Carolina State College, and Duke University. The opportunity to live for a while in a different

(handwritten margin note: Why teach in the South)

area of the nation, especially one with a relatively mild climate, might also have been a factor.

Of these 162 scientists, 40 of them began teaching with only a bachelor's degree; half of this number arrived from out of state. For instance, Norman B. Foster came to Raleigh in the autumn of 1920 to teach at North Carolina State. He had earned his bachelor's degree from Cedarville College in Cedarville, Ohio, just that year. Foster taught in Raleigh for three years, earning a master's degree at the same time. Then in 1923 he accepted a position with Women's College in Greensboro, where he remained for the next twelve years. Of the remaining nineteen people, four of them, two women who taught at Women's College and two men who taught at North Carolina State, stayed for their entire teaching careers. Six others soon left the South, eventually earned a doctoral degree, and pursued their careers in other regions of the nation. One accepted a position elsewhere without an advanced degree. Eight of these people never gained an entry in *American Men of Science* and thus left no traceable career path.[11]

Thirteen of these forty persons earned their bachelor's degrees in the South and remained to teach. One, William W. Wood, continued to teach at Davidson with no higher degree. Another Davidson faculty member, William Nelson Mebane, Jr., also taught at Davidson for his entire career, but in 1928 he earned a master's degree from Cornell University. Four others soon chose to pursue a doctoral degree. One of these men, Kelly Elmore, a native of North Carolina, remained in the South. He received his bachelor's degree from Trinity College (Duke) in 1922, taught high school for two years, then taught at Duke for two years while working on a graduate degree. Following brief stints at Tennessee State Teachers' College and Arkansas Polytechnic College, he returned to North Carolina, received the Ph.D. from Duke in 1931, and spent the remainder of his teaching career at North Carolina State. The other three southerners who pursued doctoral studies left the South for both their graduate education and their careers. The remaining seven persons disappear without a trace after teaching very briefly.

Of these thirteen people, nine of them taught at their alma

mater, indicating that southern schools were not adverse to hiring their young graduates as instructors for freshmen courses. Such was no doubt an expedient alternative to searching for and hiring more highly trained professors when enrollment figures were unpredictable. Likewise, this situation provided a good opportunity for bright young college graduates. It allowed those who were undecided about their future a chance to earn a living without committing themselves too hastily to a career path. For others, teaching freshmen for a year or two was a good way to decide whether or not to pursue an academic career. (See table 4.5.)

Fifty-nine of the 162 persons who taught in North Carolina between 1920 and 1925 began their careers with a master's degree. Of those whose careers can be traced through *American Men of Science*, 24 remained at their southern employing institution for at least ten years; 3 accepted positions at other schools, and 13 left their positions to return to graduate school and eventually earned the Ph.D. Of these 59 people, 26 came into the state having earned both degrees in another region. Only 9 of them stayed at the employing institution for at least ten years; 2 accepted positions at other colleges; 7 left their positions to earn a doctoral degree; and 8 are untraceable. None of the doctoral degrees came from a southern institution, and only 1 of these persons returned to the South, to Guilford College in Greensboro, North Carolina.

Twenty-eight of these fifty-nine people had some prior connection to the South before beginning to teach. One, Robert N.

Table 4.5
Career Path of Teachers with Bachelor's Degree
North Carolina Science Faculty, 1920–1925

Career path	No.	%
Remained at employing institution	5	12.5
Eventually earned master's degree	3	7.5
Eventually earned doctoral degree	10	25.0
Accepted other employment	1	2.5
Unknown	21	52.5
Total	40	

Wilson, was born in North Carolina but left the South to earn his bachelor's degree at Haverford College in Pennsylvania in 1898. He then taught at Guilford College in Greensboro, North Carolina, for nine years before accepting a position with the Florida State Experiment Station in 1907. While in Florida he earned a master's degree from the university and in 1910 began a long teaching career at Trinity College (Duke).

Another nine earned their bachelor's degrees in the South but left the region for a master's degree. Six of them, upon returning to the South, stayed at their employing institution for the remainder of their careers. One left after two years to earn a doctoral degree and then began to teach at Agnes Scott College in Decatur, Georgia, while two others left after six years with no trace.

Eighteen of the people teaching with master's degrees between 1920 and 1925 earned both of their degrees in the South. Two of them continued their education, earning doctoral degrees while teaching at the University of North Carolina, and then left the state, one to the University of Delaware and the other to the College of Charleston. Another three left North Carolina after a brief tenure to earn a doctoral degree elsewhere. Eight of them continued to teach in North Carolina for lengthy periods of time, and five of them soon left without a trace. (See table 4.6.)

In comparing tables 4.5 and 4.6, certain conclusions present themselves. First, during this time period southern colleges and universities were quite willing to hire instructors without a doctoral degree. Although only six of those with a bachelor's degree

Table 4.6
Career Path of Teachers with Master's Degree
North Carolina Science Faculty, 1920–1925

Career path	No.	%
Remained at employing institution	24	40.7
Eventually earned doctoral degree	13	22.0
Accepted other employment	2	3.4
Unknown	20	33.9
Total	59	

continued to teach without earning a graduate degree, twenty-six of those with a master's degree did so, for a total of 32 percent. It is also evident that a significant number of both those with a bachelor's degree and those with a master's degree, a total of 41 percent, left their teaching positions after a brief period of time and never became active enough in the scientific profession to warrant inclusion in *American Men of Science*. No doubt many of them had decided to teach for a year or two, especially if their alma mater offered them the opportunity, and then discovered that (1) teaching was not their career of choice, (2) to advance in the profession they would need additional graduate education, or (3) the rewards of a career in the private sector were more attractive. On the other hand, 23 percent of these men and women remained in the scientific profession and eventually earned a doctoral degree. For them, this opportunity to teach had cemented their attachment to a scientific career.

Fifty-five of the 162 persons who taught at one of the eight North Carolina colleges between 1920 and 1925 had already opted for a career in the sciences; these 52 men and 3 women either held an earned doctorate or, in a few instances, were working toward the degree as they taught. All but one made their way into at least one edition of *American Men of Science*. Most of them migrated into North Carolina from other regions of the nation, having already earned the Ph.D. Of those born in the South, many had chosen to leave the region for their graduate education before returning to teach. Interestingly, though, the vast majority of these 55 persons, having arrived in North Carolina, remained for their entire career.

As table 4.7 indicates, twenty-two of these men and women, or 40 percent, were born in the South (N.C. had twelve, S.C. five, Va. four, and Tenn. one). This figure is surprising only in that it is not higher. As previously noted, the growth of southern colleges created employment options that attracted scientists from around the nation, particularly from east of the Mississippi River. Nine came from the Northeast (Mass. had two, N.Y. five, Conn. one, and Vt. one), nine more hailed from the north central region (Mich. had one, Ill. three, Minn. one, and Ohio four), six were born in the mid-Atlantic states (W. Va. had one, Pa. four, and Md.

Table 4.7
Birth Region of North Carolina Science Faculty with
Doctoral Degree, 1920–1925

Region of nation	No.	%
South	22	40.0
North Central	9	16.3
Northeast	9	16.3
Mid-Atlantic	6	11.0
Midwest	3	5.5
Foreign	5	9.1
Unknown	1	1.8
Total	55	

one), three came from the Midwest (Kans. two and N.D. one), and five were foreign-born.

Not surprisingly, very few of their doctorates originated in southern institutions. The University of North Carolina awarded nine of them, and Duke one (not, however, until 1930, following its expansion from Trinity College). Of these ten degrees, seven of them went to young men born in the South and educated on the undergraduate level in southern institutions. More intriguing are the three instances of people born outside the South who earned their doctoral degrees in the region.

William Battle Cobb, although born in Massachusetts, attended the University of North Carolina for all of his postsecondary education. He completed the B.A. degree in 1912, the M.A. degree the following year, and then served as a scientist for the United States Soil Survey. From 1920 to 1924 he taught at a branch of Louisiana State University and then accepted a position at North Carolina State in Raleigh. While teaching, he began work on the doctoral degree at UNC, completing it in 1927. He remained at North Carolina State until his untimely death in 1933.

Horace D. Crockford, like Cobb, was born outside the South, in Pennsylvania, but as a boy moved to Charlotte, North Carolina, with his family. After earning the B.S. from North Carolina State in 1920, he worked for one year as a chemist with E. S.

Royster Guano Company. In 1922 he entered the graduate school at UNC and began teaching there as an instructor, earning the Ph.D. in 1926. Crockford served the chemistry department of the university for the remainder of his career, retiring in 1969.[12]

Gerald R. McCarthy, born in New York, earned his A.B. from Cornell in 1921 and then taught for one year at Williams College in Williamstown, Massachusetts. In 1922 he migrated to Chapel Hill, entering the graduate program in geology and also teaching. He earned his master's degree in 1924 and the Ph.D. two years later. Why he came to North Carolina is something of a mystery; perhaps he was familiar with Joseph Hyde Pratt, geologist for the state of North Carolina, who also taught at the university and who enjoyed something of a national reputation. Certainly the two men soon became acquainted, for during the summer Pratt hired McCarthy to work for the state geological survey. Pratt left the university to become a private consultant in 1926, the year that McCarthy completed his Ph.D. The latter then began his long career as professor of geology at the university.

Of the twenty-two scientists with a doctoral degree known to have been born in the South, fifteen of them earned their bachelor's degrees from southern colleges and then left the region to earn their doctorate. The institutions of choice were the Johns Hopkins University (six), Cornell (three), Chicago (three), Columbia (one), Rutgers (one), and Göttingen, Germany (one). All but one of them remained as professors in a North Carolina college or university for their entire careers.

Generally speaking, persons with a doctoral degree, even those born outside the South, were much more likely to remain in the region for their teaching careers than those with only a bachelor's or a master's degree. Of the ninety-nine people in the latter category who taught in a North Carolina college sometime between 1920 and 1925, only thirty-one of them, or 31 percent, remained in the state for more than ten years. Of the fifty-five people with a doctoral degree, forty-one, or 75 percent, retained their teaching positions in the state. Of course, this figure reflects the choice of those twenty-two persons born in the South, only one of whom left the state (but not the South). Additionally, Cobb, Crockford, and McCarthy, although born outside the South, had strong ties to the region and elected to remain as well. Nonetheless, of the other twenty-nine persons born and educated outside the South, sixteen of them accepted teaching positions in North Carolina and never left.

Two others, John Henry Davis, Jr., and Frederick P. Brooks, followed a somewhat different career path, although they too remained in the South. Davis, after a nine-year teaching career at Davidson College, accepted a position at Presbyterian College in South Carolina in 1930 and then in 1935 moved to Southwestern College in Memphis, Tennessee. Brooks, having earned all of his degrees from the University of North Carolina, taught chemistry there from 1923 until 1932. He then chose to enroll in the University of Michigan School of Medicine; upon completion of his degree he moved to Greenville, North Carolina, where he opened a private practice and taught some courses in health and hygiene at East Carolina Teacher's College. In 1942 he became the college physician.

Other than Davis, only twelve of those persons holding a doctoral degree chose to leave North Carolina after a few years of teaching to pursue their careers in other areas. Two of them taught in the state for only one year; one then accepted a position at another college outside the region, and the other joined a private firm as a research chemist. Three of them remained for nine years; one then accepted employment with the United States Geological Survey, and the other two moved to nonsouthern colleges. The other six stayed in North Carolina for between two and four years; five of them then went to other schools, and one accepted a position as a private research chemist.

Of these fifty-five persons with the doctoral degree, eight of them were scientists starred in *American Men of Science*: Martin Kilpatrick, Francis Preston Venable, Henry Van Peters Wilson, Alvin Sawyer Wheeler, William Chambers Coker, Otto Stuhlman, Robert Ervin Coker, and John Nathaniel Couch. In 1906 *AMS* began the rather subjective process of selecting 1,000 outstanding scientists in America from among the 4,000 persons included in that year's edition. The editor, J. McKeen Cattell, asked ten scientists from each of twelve disciplinary specialties to rank those scientists in their respective fields. While Cattell did not specify criteria, these men no doubt relied on such information as graduate degree, research, and publications to make their selections. The 1910 edition added 269 people to this list, and in 1921 the process was altered so that all men and women in a particular field had the opportunity to vote on those most prominent in their field. In each subsequent edition (1927, 1933, 1938, and 1944), 250 new names were added; following World War II, *AMS* discontinued this process.[13]

While this selection process obviously relied to a great extent on name recognition, it does afford a relative means of assessing the reputation of those scientists who taught in North Carolina. Of the eight men who were so designated, only one, Martin Kilpatrick, ever left the state. He taught at Duke from 1922 to 1925, his first position after earning the doctorate from New York University. He then spent a year at Johns Hopkins and a year in Copenhagen before settling into a teaching position at the University of Pennsylvania.

The other seven men all served the University of North Carolina at Chapel Hill and represented great diversity in region of origin, educational background, and disciplinary specialization. In the early 1920s, although Trinity College was soon to blossom into Duke University, UNC was the premier institution of higher education in North Carolina and one of the top-ranked universities in the South. A decade earlier, Governor Charles Brantley Aycock had launched the largest campaign to date to improve the state's public schools, and the university soon shared in this spirit. The legislature's annual appropriation to UNC grew from $155,000 in 1901 to more than $2 million two decades later. The first three university presidents of the twentieth century, Francis Preston Venable (1900–1914), Edward Kidder Graham (1914–18), and Harry Woodburn Chase (1919–30), oversaw rapid physical growth, encouraged creative scholarship among the faculty, and expanded graduate schools.[14]

Of no small consequence in UNC's ability to attract outstanding faculty members was its ability to pay them well, thanks particularly to the 1917 bequest of Mrs. Mary Lily Kenan Flagler Bingham. The Kenan family had long been associated with the university as students, trustees, and benefactors, although Mary Lily attended Peace Institute in Raleigh. In 1901 she married Henry M. Flagler and upon his death in 1913 inherited a sizable fortune. In November 1916 she married Judge Robert Worth Bingham, then mayor of Louisville, Kentucky, and noted newspaper publisher. Six months later she died, and her will provided $75,000 annually for a period of twenty-one years to the university to pay the salary of selected professors, who would then be known as Kenan Professors. At the end of this time, the principal of her estate would be divided among the various benefactors; eventually the university received $2.1 million to continue to fund the Kenan Professorships.[15] Thus UNC was able to offer attractive salaries and one-year research professorships with no teaching duties, making it a magnet for bright and ambitious scholars.

At this time Francis Preston Venable was nearing the end of his career, outlined in the previous chapter. A native of Virginia, he had arrived in Chapel Hill in 1880 and received his doctoral degree from Göttingen, Germany, the following year. He taught chemistry at UNC until his retirement in 1930 and served as the university's president from 1900 to 1914. He authored five books, conducted noted research on zirconium and improved the commercial manufacturing process of calcium carbide. Venable was among those first one thousand scientists nationwide whom AMS designated in 1906 as outstanding in their fields. In 1918 he was also one of the first five faculty members to receive appointment as Kenan Professor.[16]

Henry Van Peters Wilson, a biologist, was born in Maryland and received both his bachelor's degree and the Ph.D. from Johns Hopkins, completing the latter in 1888. He remained at Hopkins for one year and then worked with the United States Fish Commission at the Fisheries Biological Laboratory at Woods Hole, Massachusetts. During his two years there he completed an extensive study of the embryology of the sea bass. In 1891 he accepted the position as head of UNC's biology department, of which he was the only member. By 1904 the department had grown considerably and split; Wilson became head of the zoology department. In addition to his continuing research into the dissociation of cells and regeneration, Wilson spearheaded the founding of the United States Fisheries Biological Laboratory in Beaufort, North Carolina, an area where he had spent many summers conducting research. Along with Venable, he was chosen as one of the outstanding men in his field in 1906 and received one of the first five Kenan Professorships. He continued his active academic life until his death in 1939.[17]

Arriving to serve the chemistry department in 1900, as Venable moved to the president's office, was Alvin Sawyer Wheeler. Born in Massachusetts, Wheeler moved with his family to Iowa at a young age and attended Beloit College in Wisconsin. For three years he worked with a lumber company in Tacoma, Washington, and then taught high school there for a couple of years. Concluding that an academic career was more to his taste than lumbering, he entered Harvard University graduate school and received his doctorate in 1900. He noted in his diary that on August 8 he received word of his appointment to the University of North Carolina faculty. Evidently West Virginia University also sought to hire him and asked him to reconsider his decision. Wheeler declined and arrived in Chapel Hill on September 10; one week later, classes began. By the end of the month his wife, Edith, had become church organist, and thirty-one people had dropped by their home to welcome them to the community. In 1910 Wheeler took a one-year leave of absence for further study in Europe. Most of his research was in the area of organic chemistry and related particularly to the development of dyes. He patented Wheeler Brown, a color used in the manufacture of women's hosiery. In 1906 *American Men of Science* chose Wheeler, along with colleagues Venable and Wilson, as an outstanding scientist. Upon his retirement in 1935, Wheeler was named Kenan Professor Emeritus.[18]

The next promising young scientist to arrive in Chapel Hill was William Chambers Coker. Born in Hartsville, South Carolina, Coker attended the University of South Carolina and, following graduation in 1894, worked for two years with the Atlantic National Bank in Wilmington, North Carolina. Deciding to follow his love of nature, Coker enrolled in the graduate school at Johns Hopkins, receiving the Ph.D. in botany in 1901. He then spent a year studying in Germany and joined the faculty of UNC in September 1902. Coker became a noted expert on fungi, publishing four books and a number of articles; in addition he coauthored *The Trees of the Southeastern States* with Henry Roland Totten, a UNC colleague. He left a lasting imprint on the Chapel Hill campus by overseeing, for nearly thirty years, the university's effort to preserve the beauty of the area. He was singularly responsible for turning a swampy pasture into the five-acre Coker Arboretum. In 1934 Coker married Louise Manning Venable, daughter of Francis Preston Venable. He served the university until his retirement in 1945 as Kenan Professor of Botany, an honor bestowed on him in 1920.[19]

Otto Stuhlman, of German birth, arrived in Chapel Hill in 1920, having already established a national reputation in physics. He received his bachelor's degree from the University of Cincinnati in 1907, then enrolled in the graduate school at the University of Illinois, where he taught while earning the master's degree. He then journeyed to Princeton University, which granted him the Ph.D. in 1911. Stuhlman then taught at the Stevens Institute of Technology in Hoboken, New Jersey, for one year; at the University of Pennsylvania until 1918; at the State University of Iowa for one year; and at West Virginia University for one year before accepting the position as associate professor of physics at UNC. He served as head of the department of physics from 1928 to 1934 and remained at the university until his retirement in 1953. Although starred in *American Men of Science* in 1921, he never received a Kenan Professorship.

Robert Ervin Coker, like Stuhlman, earned a considerable professional reputation before accepting the position of professor of zoology at UNC in 1922. Born in Society Hill, South Carolina, in 1876, he received his bachelor's and master's degrees from UNC. From 1897 until 1902 he taught at a private school in South Carolina and then served as a school principal in Goldsboro, North Carolina. In 1902 he accepted the position of custodian of the newly opened United States Fisheries Biological Laboratory in

Beaufort, where he became well acquainted with Henry Wilson and served simultaneously as biologist for the North Carolina Geological Survey. Perhaps Wilson convinced Coker to continue his graduate education; from 1904 to 1906 Coker pursued doctoral studies at Johns Hopkins. Upon completing his degree, Coker worked for two years for the Peruvian government, studying that country's guano industry and marine fisheries. In 1909 he began a thirteen-year career with the United States Bureau of Fisheries, serving as a scientific assistant for one year in Washington, D.C., as director of the United States Fisheries Biological Laboratory in Fairport, Iowa, until 1915, and as chief of the bureau's Division of Scientific Inquiry until 1922. No doubt Wilson enticed Coker to come to UNC, and in 1935 Coker followed Wilson as chairman of the department of zoology. That same year he received an appointment as Kenan Professor. He introduced oceanography to UNC and worked with others to begin North Carolina's Survey of Marine Fisheries and UNC's Institute of Fisheries Research. He published numerous articles and two books before retiring in 1953.[20]

What an exciting aura the science departments at UNC must have generated in the early 1920s, with Venable and Wheeler in the chemistry laboratory, Wilson and Robert Coker anchoring zoological studies, William Coker directing the botany department, and Stuhlman continuing his work in physics. Into this milieu came John Nathaniel Couch, a young transfer student from Trinity College who in 1917 wanted to pursue a medical degree. His work with William Coker, though, turned him toward botany. Following service in the armed forces during World War I, Couch returned to the university and remained there for his entire career, earning both graduate degrees under Coker's guidance and then accepting a position as assistant professor of botany. Couch's recognition as a scientist of merit came many years later, when in 1939 the Boston Society of Natural History awarded his monograph *The Genus Septobasidium* (a group of tree-growing fungi) its prestigious Walker Prize for the most outstanding work in the field during the previous five years. He received *American Men of Science*'s designation as an outstanding scientist in 1944 and a Kenan Professorship in 1945.[21]

Obviously the University of North Carolina, with its increased legislative appropriations and especially the Kenan Professorships, was able to attract both scientists of national reputation and promising graduate students. Surrounding institutions such as North Carolina State, Wake Forest College, Women's College, Meredith College, Trinity College, and Elon College, all located within fifty miles of Chapel Hill, no doubt benefited as well. While they were unable to offer the salary scale available at the university, the proximity of a major graduate school and research library proved attractive options for a number of persons. Consequently, these schools were able to obtain many well-trained faculty members who, if not the cream of the academic crop, were nonetheless respectable professional scientists.

While a number of these faculty members remained in North Carolina only a short time, leaving either for further graduate education or a position elsewhere, a good portion of them remained in the state. Of the fifty-five persons with a doctorate who were teaching between 1920 and 1925, forty-one, or 75 percent, never left, and half of them were not southerners by birth. The greatest number of them, eighteen, served the University of North Carolina, and it is not difficult to understand their attraction to the school. However, eight taught at North Carolina State, five at Trinity College, four at Wake Forest College, two each at Davidson and Meredith colleges, and one each at Elon and Women's colleges. In addition to proximity to a major academic center, attractions no doubt included a relatively mild climate, less pressure to publish than at major universities, and the challenge of participating in a school's growth process.

Given the dedicated professionalism of those fifty-five people with earned doctorates and the national prominence of eight of them, it is not surprising to find that almost all of them were members of the American Association for the Advancement of Science and national organizations of their respective disciplines. Many of them enjoyed broad personal contacts outside of the South and had published articles in a variety of journals. Did they, then, need and support the concept of a state academy of science? Or did they agree with the critics of such organizations that the time for regional, multidisciplinary societies had passed, leaving the North Carolina Academy of Science to lesser lights and learned amateurs?

All of the eight North Carolina scientists starred in *American Men in Science* joined the academy, although Martin Kilpatrick, who taught at Trinity College (Duke) from 1922 until 1925, allowed his membership to lapse when he left the state. The other seven remained members of the academy throughout their professional lives, and each of them served a term as either president or vice-president. Additionally, they contributed a total of thirty-two articles to the journal and presented forty-three papers at annual meetings just between 1920 and 1925. Not infrequently one of these scientists teamed up with a graduate student for a presentation or shared authorship of an article, reflecting research assistance and providing the younger scholar with an opportunity to get his or her feet wet in professional circles.

While these seven outstanding scientists participated actively in the affairs of the NCAS, they by no means dominated the proceedings. Among the other forty-seven men and women who held the Ph.D. and taught in North Carolina between 1920 and 1925, thirty-five of them joined the academy, and twelve of them served one or more terms as an officer at some time before 1940. Like their more notable colleagues, they too published in the journal and presented papers at the annual meetings. Between 1920 and 1925, the *Journal of the Elisha Mitchell Scientific Society* contained 78 articles in addition to the 32 mentioned above. The programs of the annual meetings during this same time period reflect an even greater percentage of participation, for out of a total of 255 papers presented, 212 of them belonged to men and women other than those noted above.

In the final analysis, it is obvious that North Carolina scientists needed and valued the state academy of science. Nationally known men joined and participated actively in the organization and often encouraged their graduate students to do likewise. Their level of participation indicates their sense that the NCAS was a creditable professional forum. At the same time, educators without prominent reputations, even those without advanced degrees, felt equally comfortable presenting papers at meetings and publishing research results in the journal.

While level of participation is a good indicator of a thriving organization that is vital to the professional lives of its members, other, less tangible benefits should not be overlooked, although they are difficult to measure. For instance, how many of these men and women with advanced degrees remained in North Carolina colleges, educating the state's young people and encouraging them to strive for excellence, because of the contacts provided through the NCAS? During the early 1920s, it was one of the most active state academies in the South. Perhaps the proximity of a vibrant scientific society outweighed factors that might have pulled competent faculty to other institutions.

Then too, participation in academy activities could have had considerable impact on graduate students and on younger faculty with bachelor's and master's degrees. Perhaps rubbing elbows with more advanced colleagues inspired some of them to pursue a doctoral degree or to further their education on a less formal basis. At the least, such contact offered them a greater sense of professionalism and a confidence in themselves and in the need to educate young men and women concerning the latest scientific advances and theories.

Whatever their individual reasons, 206 persons had joined the NCAS by 1924. By 1930 membership had climbed to 305 persons. Other academies' memberships were on the rise as well; in 1930, the Alabama academy had 165 members, the Virginia academy had 535 members, and the Tennessee academy had 420 members. Slower-growing academies included that in Georgia, which with its restrictive membership policy had only 59 members; the one in South Carolina, with 60 members; and the one in Louisiana, also with fewer than 100 members. Florida, Mississippi, and Arkansas would not have state academies of science until a few years later. Despite the unevenness of growth in the state academies across the South, their strength in several states and eventual emergence in others indicates that southern scientists saw potential in being a part of such organizations and labored to make them viable and creditable societies.

5

Camaraderie, Research, and Publication: State Academies and Scientists' Professional Needs

During the 1920s and the 1930s, many of the southern state academies of science aspired to serve regional scientists by replicating the functions of national organizations. The state academies offered annual meetings that most anyone could afford to attend. Given the scant time and funds allocated for research by most southern colleges and universities, the academies encouraged *modest* research projects and tried valiantly to publish journals so that such efforts would not go unrewarded. Thus many southern scientists turned to the state academies of science, hoping that the organizations could ameliorate their sense of frustration and enhance their professional self-image.

The academies met these needs with varying degrees of success. While they all sponsored annual meetings, only the academies of Tennessee and Virginia found ways to support members' research endeavors in any meaningful fashion, and prior to World War II only those of Tennessee and North Carolina published creditable journals. For the most part, the southern academies of science found themselves coming of age in a difficult time, given the harsh economic realities of the 1930s. Just as many of them matured to the point where they might be able to offer the professional support so much needed by southern scientists, World War II demanded the attention of the membership, and academies drifted for a few years. Nonetheless, although none of the academies was able to fulfill completely the professional needs of its members, these organizations were important to southern scientists at a time when they had few alternatives.

Annual Meetings

Of primary concern to scientists everywhere was an opportunity to meet together; hence all of the state academies sponsored annual meetings that convened at different points around the state. Various colleges and universities agreed to host the annual events, which most often meant providing auditorium space for one or two days and occasionally one evening meal. Scientists either stayed with their colleagues in the city or lodged in a local hotel. Morning and afternoon sessions focused on the presentation of research papers. Usually the members gathered for a luncheon that was either preceded or followed by a short business meeting. As these annual gatherings became more widely publicized, various industries, particularly manufacturers of scientific equipment, asked for and received permission to display their latest products.

Unfortunately, most academies maintained attendance records only sporadically, rendering a year-by-year comparison impossible. However, the figures available do indicate that southern scientists were hungry for such contact with their colleagues. The third annual meeting of the Virginia academy, held in 1925, attracted 107 registrants, with 84 of the 237 academy members present. Eleven years later 408 people registered for the two-day convention, including 240 of the 695 academy members. Other state academies experienced a similar pattern of attendance. In 1923 the annual meeting of the North Carolina academy attracted 92 of its 201 members. The Alabama academy, whose membership grew from 165 persons in 1930 to 256 in 1940, counted between 68 and 125 members at the annual meetings, with visitors increasing the total attendance to between 100 and 164 persons. In 1935 the South Carolina academy's annual meeting attracted approximately 200 people, although officially the membership stood at just 166 persons. The next year, however, the membership figure had risen to 234 persons, and it continued to grow.[1]

While those in attendance enjoyed the companionship made possible through the annual meetings, they were interested in more than socializing. They came primarily to present the fruits of their research to an audience of their peers and to hear the same from others. In the early years of the state academies, all papers were presented to the entire assembly. The difficulty of this arrangement, however, soon became obvious. Although biologists listened patiently to chemical analyses and chemists offered the same courtesy to the botanists, many members thought that sectional divisions would prove more beneficial to speaker and audience alike.

The growing number of papers submitted for presentation further complicated the scheduling of sessions. The program of the first annual meeting of the Alabama academy (1924) contained 17 papers. Within five years the number had expanded to 28 papers, and in 1933 some 70 papers were scheduled to be read. The Virginia academy's inaugural meeting in 1923 included 18 papers. Six years later the program listed a total of 62 papers, a figure that leaped to 108 within two more years. Although the early meetings of the North Carolina academy contained fewer than 15 papers, this figure never dropped below 23 after 1909. By 1923 the program listed 59 papers; this figure fluctuated between 50 and 80 until the eve of World War II. The Tennessee academy, which held meetings semiannually, saw an increase in the total number of papers from 25 in 1926 to 53 by 1930, 67 by 1935, and 61 in 1940. Even the relatively static South Carolina academy saw its annual program expand from 15 papers in 1925 to 21 in 1930, 33 in 1935, and 45 in 1939.[2] Consequently, well before 1940 all of the academies, even those with modest attendance, offered concurrent interest sessions at their annual meetings.

They maintained as well at least one general session for papers that could not readily be categorized or for those that had been specifically solicited for presentation to the entire membership. These papers ran the gamut from specialized research reports to matters of broad public concern such as conservation and education. Occasionally the conferees were treated to a symposium in which several members addressed a specific topic from their varied perspectives. The Tennessee academy, which was the only one in the South to sponsor two meetings each year, frequently offered such a program at its spring session. Topics included the resources of the state, caves of Tennessee, biology of the Great Smoky Mountains National Park, and implications of the Tennessee Valley Authority for the economy and resources of the region.

Annual meetings also provided for scientists secondary contact with their peers in other states. By 1926, eighteen state academies of science throughout the nation had established formal affiliation with the AAAS. In that year, J. T. McGill of the Tennessee academy suggested that each academy send at least one representative to the annual AAAS meeting and that these delegates form an academy conference to discuss informally matters pertinent to all of the state academies. During subsequent years frequent topics of discussion included increasing membership, a journal exchange program, sources of financial support, improving the science education in secondary schools, and finding the time and means for research.[3] The delegates then carried these discussions back to their state academies, effectively knitting together the interests of otherwise disparate regional organizations. As will be noted, ideas shared in Academy Conference meetings sometimes spurred action in academies across the nation.

Journals and Other Publications

Some of the research reports presented at the annual meetings were significant enough for publication in national professional journals. State academies took pride in these accomplishments, and when their journals printed abstracts of such papers, or even occasionally the entire paper, they noted the national journal in which the work was to appear. Unfortunately for the academies, scientists preferred to submit their work to national journals, relying on the state publications only when more prestigious ones declined their articles. In 1927 J. T. McGill, the Tennessee academy's longtime secretary, noted that relatively few of the papers presented before the academy since 1912 had been published in the organization's *Transactions* because "the authors preferred to publish in the special journal of the science under which the research was classified."[4] While southern scientists should not be censured for seeking the widest possible audience for their research, which by the 1920s had become an important component of their professional career, their lack of confidence in the state journals did not help the latter's cause.

All of the state academies recognized this need of publication, and most of them aspired to publish creditable journals. Prior to World War II, though, only a few succeeded. Most of the young organizations managed only small issues that included abstracts of papers presented at the annual meetings and secretaries' reports. Much of the problem was, of course, financial; scientific papers frequently included drawings, graphs, and tables that required a production process beyond the limited means of these organizations. Then too, since these journals seldom attracted the papers of top-notch scientists, they could do little to broaden their circulation beyond the academy membership and its journal exchange program. Eventually some of the academies sought industrial sponsors and advertisers, but the small amounts of money thus raised made little appreciable difference.

Among the southern state academies of science, the journals of North Carolina and Tennessee were by far the strongest. North Carolina scientists were extremely fortunate to have two organizations, the Mitchell Society as well as the academy, to share the expense of a journal. Although the early issues of the *Journal of the Elisha Mitchell Scientific Society* were small and appeared irregularly, the publication had become a respected periodical by 1920. The volume for 1919–20 contained thirteen articles as well as the proceedings of the North Carolina academy. Eight years later the 256-page *Journal* published a total of thirty-two articles plus the proceedings of the academy. By 1935 the *Journal* had expanded to 375 pages and illustrations.

In neighboring Tennessee, the state academy established a journal that, after a rocky financial beginning (only two issues appeared between 1914 and 1925), was every bit as professional as that of North Carolina. The modest first issues of the Tennessee academy's *Journal* averaged twenty to thirty pages and included two or three papers, the proceedings of the academy, announcements, and a membership list. In 1930 a somewhat discouraged editor expressed his desire for a publication that could include more and longer articles as well as the pictures and illustrations often required by scientific writings. Complaining of the sluggish cash flow that kept the *Journal* months behind schedule, he suggested that members solicit their friends to join the academy and that they also consider becoming sustaining members by contributing ten dollars per year instead of paying only the regular two-dollar membership fee. One year later the executive committee authorized the editor to accept appropriate advertising, and shortly thereafter the *Journal* began to expand. The four issues published in 1933 contained 436 pages and included thirty-eight complete articles as well as the proceedings of the academy, announcements, and a membership list.[5]

The articles published in these two journals reveal a broad spectrum of interests among scientists, as well as an awareness of national and worldwide scientific discussions and a concern for their community and region. Some articles, of course, offered little more than a review of developments in a specific field, such as Andrew H. Patterson's "The Theory of Relativity," published in the 1920–21 issue of the *JEMSS*; J. L. Lake's "The Search for the Ultimate Atom," in the 1922–23 issue of the *JEMSS*; and Francis Preston Venable's account of the work of Robert Brown, "The Brownian Movements," published in the 1924 issue of the same journal. While such articles were not likely candidates for national journals, they did meet several of the purposes of state academies as suggested by W. S. Bayley and Paul Boyd in their defense of the organizations. As noted in the previous chapter, Bayley maintained that one of their concerns should be education of the general public, and Boyd insisted that the academies should act to offset the myopia of increased specialization.[6] Then too, many younger educators, especially those without a specialized graduate degree, no doubt found such review articles of great value.

On the other hand, a number of articles moved beyond the repetition of theories and research of renowned scholars to include original commentary. Venable published a number of reports detailing his experiments with various chemical elements, culminating with his 1922 article "Isotopes" (chemical elements that differ so slightly from one another due to radioactive discharge that they are not considered separate elements at all). Venable concluded that such elements were heterogeneous physically but homogeneous chemically. "The masses of all elements," he maintained, "are composed of chemically similar particles. So long as this is true the chemist at least need not disturb himself." Of course, wrote Venable presciently, "there always looms before us the fact of atomic disintegration."[7]

Many articles published in these early journals concerned regional matters, in large part because research conducted by southern scientists that deserved national coverage usually found an outlet through one of the national journals, but also because many southern scientists focused their attention on issues of local concern. Consequently, articles describing the flora, fauna, and mineral deposits of specific areas abounded. Frequently scientists applied their expertise to such issues as the development of the Tennessee Valley Authority and local agricultural problems. The journals also provided a forum for scientists to speak out on matters of social concern. In a series of articles Tennessee scientists of the 1920s expressed their dismay over the state law banning the teaching of evolution in public schools. Scientists in many other states authored a number of articles outlining various suggestions for improving secondary science instruction.

The Alabama academy realized less success in its attempt to publish a journal than did the academies of North Carolina and Tennessee. In 1926, three years after its organization, the academy printed a volume of abstracts of papers presented at the first three meetings. Funding for an annual journal proved elusive, however, for in 1928 when the membership voted to publish abstracts "at intervals when funds and materials were sufficient," the academy's bank balance was a mere $30.28. In 1930, with an expanded treasury of $123, the body decided to forge ahead and publish an annual edition of "Abstracts and Proceedings." Born like many of the other journals in an economically unstable decade, the *Journal of the Alabama Academy of Science* appeared only irregularly throughout the 1930s and did not begin to include complete papers until 1943.[8]

The journals of the other southern academies of science reflected the experience of Alabama. South Carolina published nothing until the inauguration of its *Bulletin* in 1935, which contained only proceedings of the annual business meeting and abstracts of papers read. In 1940 the membership voted to raise sufficient funds to expand the *Bulletin* to include some complete papers, but World War II so disrupted the academy that the matter was temporarily set aside; not until 1957 did the *Bulletin* print complete papers.[9] The Louisiana academy in 1932 published the first volume of its *Proceedings*, which did contain fourteen complete papers. The journal did not appear regularly, however, until 1949, and even then some issues contained only minutes, officers' reports, and membership lists.[10] The first issue of the Georgia academy's *Bulletin* appeared in 1935 with the program of the annual meeting and abstracts of papers. Although the executive council noted in 1936 that University of Georiga president Harmon W. Caldwell favored a university subsidy for the publication, an agreement beween the two parties never materialized. Not until 1953 did the *Bulletin* publish complete papers.[11]

The Virginia Academy of Science followed a somewhat different path in its publication efforts. The academy annually printed its proceedings but did not establish the *Virginia Journal of Science* until 1940. After only two years financial stringencies compelled the academy to suspend publication of the *Journal* until 1950. The Virginia academy was, however, successful in producing other types of publications. In 1930 it published *The Flora of Richmond and Vicinity*, made possible by a grant of one thousand dollars from the state legislature.[12] Beginning in 1933, the Flora Committee issued a bimonthly mimeographed pamphelt, *Claytonia*, essentially a newsletter on wildflowers. *Claytonia* survived until April 1939, after which it was supplanted by the sometime *Virginia Journal of Science*. During the 1930s the academy also initiated an extensive study of the James River Basin that addressed industrial possibilities of the region as well as a survey of the basin's natural history. The turmoil of the war years delayed the planned publication of the study, however, and it did not appear in print until 1950.[13]

The academies' inability to publish creditable scientific journals stemmed, in part, from perpetual insolvency. With annual dues that averaged two dollars per person and no other source of income, these organizations could hardly afford the expense of the annual meeting, let alone consider the publication of an annual or quarterly journal. Then too, they experienced difficulty attracting top-notch articles, for scientists with research worthy of national attention usually submitted their work first to national journals. Consequently, these journals attracted little attention outside of the academy membership. By and large, the articles that appeared in the publications of the southern state academies of science either focused on subjects that were peculiar to the region or addressed narrow topics that were not of interest to national journals.

While the state academies for the most part proved unable to realize the publishing aspirations of their members, they did not fail completely. Meager as most of these journals were, the state academies used them to establish exchange programs with other scientific societies throughout the nation. In such a manner, southern scientists found at their disposal a wide variety of journals that they would otherwise have been without. Moreover, the southern journals, even the abbreviated bulletins, broadcast that scientific inquiry was alive and well in the region and afforded some indication of the various research projects then underway. Such publicity was far more than southern scientists would have enjoyed otherwise.

Research: The Pressure of Time

A third major concern of pre–World War II scientists, in addition to interpersonal contact and a publication outlet, was securing adequate support for research endeavors. The founders of the various state academies intended to encourage one another in the pursuit of research. The initial constitution of the Tennessee academy listed as its first objective "scientific research and the diffusion of knowledge concerning the various departments of science." The only purpose listed in the Alabama academy's first constitution was "the study and the advancement of science."[14] Expressed more specifically, scientists hoped that their contact with one another through the academies would overcome the intellectual isolation that was so detrimental to the furtherance of original research. They assumed that such organizations, by offering the opportunity to present the fruits of research and to meet with colleagues, would provide a necessary stimulus to research activities. Their enthusiasm was soon dampened, though, by the financial realities of most southern colleges and universities, which dictated overworked faculty members and a paucity of time and money for research.

Time away from classroom duties was, of course, a key element for scholarly inquiry. Other than some of the Kenan Professors at the University of North Carolina, few southern academic scientists enjoyed the luxury of an appointment that specified a certain proportion of their compensated time for research. Rather, they taught three or four courses each semester in their various specialities and supervised the laboratory work of their students as well. The assignment of graduate assistants for such duty had not yet begun, for only a handful of southern colleges even offered graduate degrees, and those that did had few funds to subsidize their students.

While southern science professors accepted an economic situation that left them with full teaching loads, they expressed dissatisfaction with the policy of the Southern Association of Colleges and Secondary Schools that allowed administrators to count two hours of laboratory instruction as only one classroom hour. As early as 1925 Donald Davis, a professor at William and Mary College, presented the Virginia academy with a motion requesting that the Southern Association amend this recommendation so that laboratory and classroom hours would be considered equivalent. Although the motion was tabled at this meeting, it was reconsidered and passed in 1932. The secretary of the Virginia academy, E. C. L. Miller, took this matter to the national meeting of the AAAS in 1933, presenting a paper that appeared later in the official AAAS proceedings, which the Virginia academy then distributed to all of the state academies of science, hoping for some concerted action.[15] In 1934 the North Carolina academy adopted a similar resolution and concluded that professors should spend no more than fifteen or sixteen hours per week total in instruction. Ten years later the Alabama academy approved a similar resolution, and the Georgia academy did likewise in 1946.[16]

Although the academies expressed genuine alarm over this issue, it would seem that they exerted little influence over either the Southern Association or college administrators. Resolutions went no further than printed proceedings and proved singularly ineffective as a way of bringing about change. Excepting the action of the Virginia academy, no evidence exists that academy committees forwarded their resolutions either to the Southern Association or to college administrators.

This problem of a lethargic membership, such as noted earlier by David Whitney of the Nebraska academy, would plague state academies of science from their inception to the present and explains in part why the state academies never became strong organizations. Usually, a handful of officers and committee chairpersons handled the societies' business between meetings. Members willingly turned out for the annual sessions, glad of an opportunity to greet their colleagues and present an occasional paper, but few assumed responsibility otherwise. The lament of Patrick Yancey to R. S. Poor, both active members of the Alabama academy, could apply to any of the southern academies of science: "It is strange how after the meetings people seem to drop everything concerning the Academy and that only just before the next meeting do they show renewed interest."[17]

Research: The Tennessee Academy Initiates a Program

Reelfoot Lake an exception!

Lethargy among academy members was apparently not a problem, though, when scientists in Tennessee adopted a project. In 1925 Scott C. Lyon, professor of biology at Southwestern Presbyterian College in Clarksville and president of the Tennessee academy, turned his colleagues' conservation interests into a potential for research when he suggested the establishment of a biological research station at Reelfoot Lake. Located in the northwest corner of the state, the lake, a mixture of swamp, marsh, and open water created by an earthquake in 1812, had remained a relatively undisturbed habitat for a variety of flora and fauna. Invaded only by an occasional fisherman and a few adventuresome biologists, the area surrounding the lake was, according to Lyon, an ideal mecca not only for biologists but also for organic chemists, geologists, and meteorologists.[18]

Maintaining that "large things grow safer from small beginnings," Lyon suggested that the academy seek state assistance for the project by asking for a suitable site near the lake on government-owned land. Once a location had been secured, the members could construct a "camping shack" to house both scientific paraphernalia and transient investigators. All accredited scientists, including students, would be welcome to utilize the facility. Lyon figured that the initial expenditure would require no more than a few hundred dollars, which could perhaps be raised from "friends of science." He also dreamed that perhaps "the State might be interested more than in a merely passive way."[19]

Hoping to cash in on Governor Austin Peay's campaign promise of the year before to purchase land for parks both at Reelfoot Lake and in the Great Smoky Mountains, Lyon encouraged the Tennessee academy to spur him to action with a reminder in the form of a resolution. As usual, the wheels of government turned slowly, but by 1931 the state had acquired a considerable amount of property around the lake. In that year, the legislature approved the academy's request for a biological station in the area, instructed the Game and Fish Commission to set aside an existing building and ten acres for the use of the academy, and, uncharacteristically for parsimonious southern state legislatures at the time, threw in $2,500 "for expenses."[20]

Upon inspection, Tennessee scientists discovered that Walnut Log Lodge required a few minor repairs. Rotting steps, lack of screens, a leaky roof, the absence of running water and electricity, and no kitchen equipment rendered the cabin a less than desirable summer residence. A number of academy members volunteered to oversee different aspects of the renovation, and within a year the building was habitable, scientific equipment had been installed, and boats were at the dock, ready for use. In August 1932, forty-four members of the academy assembled at the station to dedicate the John T. McGill Laboratory Building, a fitting tribute to the

Vanderbilt University chemistry professor and secretary of the academy who had overseen most of the project.[21]

Throughout the 1930s the Tennessee academy expanded its work at the Reelfoot Biological Station. The latchkey remained out to individual researchers and classes from academic institutions in the state. Summer seminars attracted faculty and graduate students from across the South, and members of the academy assembled there for both formal and informal meetings. The state legislature continued its financial support by appropriating $2,500 in both 1933 and 1935 and $5,000 in 1937. Modest fees charged to those who used the facilities met additional expenses. As an incentive to research, the academy offered a number of "scholarships" to members who proposed studies concerning the preservation and ecology of the area. In 1939 one issue of the *Journal* featured papers generated by work at Reelfoot Lake, and single papers appeared in almost every issue.[22]

Research: Funding

The Tennessee academy was the only one in the South to maintain such a research facility. Scientists elsewhere, left to their own devices, encountered financial difficulties that often proved more of an impediment than their heavy teaching loads. For the most part the academies were unable to alleviate this predicament, although some very modest funding came to them for distribution from the American Association for the Advancement of Science. Beginning in 1920, the AAAS, hoping to increase its membership, offered to various regional scientific organizations a fifty-cent payment for each of their members who also joined the AAAS. Originally this money carried no stipulations as to its use, but in 1934 the national organization suggested that local societies earmark the funds for research grants.[23] Realizing that many state academies had become accustomed to using the funds in their general budgets, the AAAS encouraged them to accept the change by increasing the amount awarded. Instead of paying a flat rate of fifty cents for every common member, the AAAS voted to award the money in twenty-five-dollar increments that would continue to be based on common membership. Thus, under the old plan the Alabama academy had received sixteen dollars in 1934, but it received twenty-five dollars the following year.[24]

The AAAS was not wrong when it anticipated some resistance to their change of policy. Many members of the Alabama academy, for instance, wanted to continue the old refund procedure. President A. G. Overton, letting it be known that he had "made [other] arrangements that the refund of $16.00 will be taken care of," hoped to win the membership over to the idea of establishing a research fund. In actual fact, Overton contributed the sixteen dollars from his own pocket, and within a year the Alabama academy awarded the research grant of twenty-five dollars, quite meager even in the 1930s, on an annual basis.[25]

Other academies adapted more readily to the change. The members of the South Carolina academy voted to accept the money as a research grant instead of using it for general expenses. So did the Tennessee academy, whose large membership entitled it to a fifty-dollar allotment. Members of the North Carolina academy not only supported the idea but began immediately to discuss the proper method of distributing the funds. They concluded that at least initially the award should go to someone engaged in research at one of the state's smaller colleges who did not have access to even the limited funds and equipment available at the major universities. In a newsletter circulated in November 1935, the academy outlined specific provisions for awarding its seventy-five-dollar stipend.[26]

Interestingly, although the academies formed grant-in-aid committees and notified their memberships of the available funds, they received few applications for the awards. Certainly the stipends were not large, but they could purchase a new piece of equipment or several books. They could also fund a few days' travel for hardpressed scientists struggling to maintain their professionalism during the Great Depression. Yet of the 226 members of the North Carolina academy in 1935, only 5 submitted applications for the seventy-five-dollar award.[27] In 1937 the academy's committee on research grants received only one application, despite the distribution of a special circular. Donovan S. Correll, a Duke University graduate student, requested travel funds to continue his study of orchids. One of the members of the grant committee, O. J. Thies, professor of chemistry at Davidson College and an unsuccessful candidate for a grant the previous year, was less than delighted with the single prospect. Writing to W. C. Coker, chairman of the committee, he stated that "I am rather opposed to granting an award when there has been no competition. Also, I had thought of the award going into something a little more tangible than 'traveling expenses.' . . . 'Traveling expenses' may mean a nice vacation, or may mean a lot of sincere hard work." Admitting that he did not know Correll personally, Thies grudgingly voted to award him the money.[28]

The records of the other academies of science reflect a similar lack of interest in the AAAS-funded research grants. In 1936 the South Carolina academy decided to follow the example of the North Carolina academy and grant the awards to faculty members whose institutions offered no support for research. Yet one year later the committee complained that "very little interest is being shown in the Research Grant. The 1936 Council voted to give preference to applicants from schools not supporting research. No applications have been received from these individuals." Two years later the committee repeated that "we need greater interest shown in this grant, and especially from that portion of the membership connected with the smaller institutions." As it was, the first four grants went to faculty members at the Medical College of Charleston. Evidently the research committee preferred to choose from among the small number of applicants rather than to decline the funds.[29]

The situation in Alabama was no different. The first award in 1936 went to Septima Smith, a University of Alabama zoologist and the only person to apply for the funds. She won the award again in 1937. In 1938 her research associate, J. Gordon Carlson, received the award. J. Allen Tower of Birmingham-Southern College accepted the grant for both 1939 and 1940. Although the academy records do not always mention the exact number of applicants for the award, they do indicate a low interest; only two persons submitted a request for the funds in 1940.[30]

Scientists in Tennessee showed only slightly more interest in the grant, which for that academy began with fifty dollars and increased to seventy-five dollars in 1939. From twelve applicants in 1937, the research committee chose to award the fifty dollars to Horace B. Huddle, professor of chemistry at the State Teacher's College in Johnson City, for his "Study of the Oil of the Tennessee Red Cedar." Other pre–World War II recipients included Hiram M. Showalter, whose name does not appear on any membership list; Glen Gentry, a high school teacher; Dorr Bartoo, professor of biology at Tennessee Polytechnic Institute; and Edward McCrady, Jr., professor of biology at the University of the South.[31]

The Louisiana academy's initial allotment from the AAAS was only twenty-five dollars. The academy had high hopes of inaugurating a major research grant when in 1938 President Harold L. Kearney received the offer of a five-thousand-dollar bequest from a New Orleans resident. Upon being informed by an attorney that the academy should be incorporated before accepting such a bequest, Kearney immediately petitioned the state for a certificate of incorporation. The certificate was issued in January 1939, but by this time the benefactor, for reasons unexplained, had withdrawn his offer. Other than the small AAAS grants, the only research awards offered by the Louisiana academy were medals for outstanding papers at the annual meetings.[32]

Of all the southern state academies of science, only Virginia undertook the task of raising an independent research fund. In 1925 academy president J. Shelton Horsley appointed a committee "to concern itself with the advancement of scientific research in Virginia." Like its counterparts in other state academies, the committee had no real idea of what it could possibly do beyond perhaps offering some small recognition for an outstanding paper presented at the annual meeting. Horsley saw this proposal as no solution and suggested "that it will be a good idea to raise about $25,000 to be at the disposal of the Committee on Research of the Virginia Academy of Science." He figured that the interest on such an amount would be about $1,200 annually and could be divided between the necessary expenses of the committee's work, a cash award for a meritorious paper, and a research grant for a scientist in need of financial assistance.[33] As far as many academy members were concerned, Horsley might as well have suggested a summer vacation on the moon.

Undaunted, the wealthy Richmond surgeon seeded the fund with a donation of $150 to be used during the three succeeding years as an award for an outstanding paper at the annual meeting. Then he turned his attention to raising funds for the research endowment. By March 4, 1927, he reported to an incredulous membership that the special account contained slightly over eight thousand dollars. Three individuals, including Horsley, had contributed one thousand dollars each, four persons had offered five hundred dollars each, and numerous others, most from the Richmond area, had contributed between one hundred and two hundred dollars.[34]

Although Horsley had overestimated the amount of money he could raise, the research committee in 1929 notified the membership that it would begin to accept requests for grants. In its first year of operation, the research fund contributed a total of four hundred dollars to six different projects. Between 1929 and 1940, seventy-one persons received grants totaling about forty-five hundred dollars. Of course, not all requests were honored, and frequently only part of the amount requested was awarded. In some instances the research committee attached stipulations to the grants. In 1931, for instance, it agreed to honor the one-hundred-dollar request of Thomas W. Turner of Hampton Institute only if the institute would match the amount. In 1935 two awards carried such a stipulation. Sometimes the committee received requests that were too large for its limited resources, but in the interest of encouraging research in any way possible, it usually referred such requests to foundations with more extensive resources.[35]

This program did not operate without difficulty. Problems of record keeping and deadlines sometimes created ill will among academy members. E. C. L. Miller, the secretary of the academy, grew a bit testy in 1931 when he learned that the committee had rejected two applications because they had been filed a few days late. "Now if the purpose of this Committee is the encouragement of research in Virginia," he chided committee chairman Horsley, "then we should be very careful not to discourage it. The spirit of research is so scarce, so difficult to arouse and so easily extinguished that we should not throw cold water on any little flame we find."[36]

Such a strict policy on the part of the committee is surprising, given its contention that, even with its relatively large fund, too few members applied for grants. Between 1935 and 1940 the committee received fifty-four applications and honored thirty-seven of them, at least in part. By 1940 the committee's report to the membership sounded much like those of other academies as it pleaded for more requests, particularly from members in some of the state's smaller institutions. Of the seventy-one initial grants, forty-two went to faculty members at Virginia's three largest schools—the University of Virginia, the Medical College of Virginia, and Virginia Polytechnic Institute. Hollins and Bridgewater colleges received one each, and the remaining twenty-seven were divided among the University of Richmond, William and Mary, Emory and Henry, Hampton Institute, Virginia Military Institute, Randolph-Macon College, and Lynchburg College. Many people applied for and received more than one grant, reducing even further the spread of funds among the total membership.[37]

Despite such difficulties, the research committee constantly strove to increase the endowment fund. When the AAAS initiated its research grant program, the large Virginia academy became eligible for one hundred dollars annually. Horsley also continued his fund-raising efforts. He enlisted the support of the pens of Douglas Southall Freeman, a distinguished historian, and Virginius Dabney, editor of the *Richmond Times Dispatch*.[38] He also issued a patriotic pitch to a number of individuals whom he thought might donate to the fund, including Mrs. Alfred duPont, a native Virginian then residing in New Jersey. Outlining how the monies had been utilized so far, Horsley appealed to their love of the South by claiming that without an enlarged research fund, Virginia could not hope to compete with the better-endowed institutions in other sections of the nation. "If we could keep only one out of four research workers and well-trained scientists in Virginia, who are now lost to us," he concluded, "we would be doing something very great for the state." Horsley then added, "This work naturally does not have the emotional appeal that crippled children, illness or the indigent have, and so it must be limited to a few persons of understanding, who give intelligently rather than emotionally, and these must be asked to give in larger amounts." Mrs. duPont sent a check for one thousand dollars, and others contributed lesser amounts. By 1940 the endowment had grown to slightly over thirteen thousand dollars.[39]

Some of the academy grant committees expressed concern about the way in which the funds were utilized. A number of persons, like Thies of North Carolina, feared that grants for intangible purposes could easily be misused. In 1939 E. C. Faust of the New Orleans Academy of Science presented a report at the annual AAAS meeting in which he outlined how the grants had been used to date. Of a total of 175 grants made through every state academy of science (not just those in the South), only 52 of the recipients had produced papers. Faust urged that state academies award grants only for definite research projects and that the academies insist on some sort of report indicating the use of the money and the results of the research.[40] A few of the academies took such action; most of them, though, while encouraging grant recipients to publish their findings, never required a formal accounting.

Two other efforts to support research among southern scientists are worth noting. In 1936 Phipps and Bird, Inc., a Richmond, Virginia, manufacturer of scientific instruments, offered an annual prize of a gold medal for an outstanding paper presented before each of the academies of Virginia, North Carolina, South Carolina, and Georgia. In addition, the winning paper from each academy would be submitted to a panel of judges chosen from among the state academies to compete for a first prize of one hundred dollars and two additional prizes of twenty-five dollars. Lloyd C. Bird, president of the company and an active member of the Virginia academy, hoped that the prizes would elicit some top-quality competition. All four academies accepted this offer and turned the matter over to their respective research committees. Until 1943, when Phipps and Bird discontinued the program, academy members competed annually for the Jefferson Gold Medal, although once again research committees complained of the lack of entries. Evidently Bird was also disgruntled, for in announcing the discontinuance of the award he stated that it "had not accomplished the purpose for which it was designed."[41]

Several of the southern academies' research funds also benefited somewhat from the curious generosity of C. M. Goethe, a banker from Sacramento, California. The banker's initial offer was made to the Virginia Academy of Science in 1943 after he read of the academy's research fund in the alumnae magazine of Randolph-Macon Women's College, from which his wife had graduated. Within a year both Goethe and his wife agreed to donate two hundred dollars annually to the research fund if the academy would raise an additional four hundred dollars.[42] Evidently Goethe, a member of Sacramento's city planning board, had become disillusioned with the "men absorbed in profit-making" whose inertia effectively blocked such goals as city parks. "It is for this reason," he continued, "that I feel the extreme need of doing everything possible any individual can toward conservation of pure research." Two months later in a letter to E. V. Jones of the Alabama academy, he added:

Being stockholder in several score corporations, I have noted the complete change from an attitude of sneering at research say 20 years ago, to the present one where some corporations spend hundreds of thousands of dollars in research.

This is all good. However, do you not sense an element of danger in discovery being owned by those whose main object is a profit. Might this not result in even the sidetracking of a discovery that might effectually injure the ledger account?

Does it not seem important to keep alive the spirit of pure research? This is probably best accomplished by Academies of Science.[43]

Yet Goethe's motivation was not so simple as it first appeared. While he always insisted that he placed no restrictions on the grants, he noted at least once that "we do hope that, if not now, then at some future time it will be expended toward research in either human genetics or eugenics." Several months later Goethe revealed that he had founded the Eugenics Society of Northern California and provided its entire support, which amounted to approximately five thousand dollars in 1943.[44]

Other academies of science soon learned of Goethe's offer to the Virginia academy, and by the late 1940s those of Alabama, North Carolina, South Carolina, Georgia, and Florida had also secured contributions from the eccentric philanthropist. Eventually the seeming windfall proved to be more trouble than it was worth. The contributions, scheduled to arrive quarterly, seldom did. Matters were further complicated by the decision of Goethe and his wife to send separate checks. Finally, Goethe proved unwilling to make longterm commitments concerning the contributions. He agreed to the arrangements only on an annual basis, indicating that he hoped he and his wife could continue to make the bequests for the remainder of their lives. While the Goethes are mentioned infrequently in a number of academy proceedings throughout the 1950s, it is difficult to determine just how much they actually did contribute to various research funds.[45]

Despite contributions from the AAAS, Phipps and Bird, and the Goethes, academy research funds remained almost miniscule. With the exception of the Virginia academy, which could afford to make several grants annually of up to one hundred dollars each, most of the awards amounted to only twenty-five or fifty dollars. The example set by the Virginia academy shows that with some serious fund-raising, at least modest goals could be reached. However, such effort required dedicated individuals like Shelton Horsley; few southern scientists had such time to spare or the extensive personal contacts that Horsley enjoyed. For the most part, then, the academies parceled out whatever funds came their way but did little to raise more.

Even though the awards were small, the apparent lack of interest in these funds on the part of academy members is perplexing. All of the committees bemoaned the small number of applications received. They especially lamented the general failure of scientists in the small institutions to take advantage of the grants. The academies of North and South Carolina specifically stated that priority would be given to such applications, with few apparent results.

Several factors might account for this poor response. First, few persons teaching at smaller institutions applied for the grants because, holding only a master's or bachelor's degree, they had been educated more as teachers than as scientists. As E. C. L. Miller of the Virginia academy pointed out, "The training that teachers receive frequently does not develop a strong desire to do research work and when difficulties arise the research work is more and more neglected and finally abandoned."[46] In short, the job of teaching at a small college did not require productive research, nor did these instructors often aspire to a more prestigious position in the scientific community.

Additionally, many young educators whose interests did include research left their employment to return to graduate school. While young scientists in North Carolina could, if they chose, earn a doctorate at either the state university or Duke University and simultaneously retain their teaching position, those in other states (with the exception of Virginia and Texas) did not have the luxury of a nearby graduate school. Having earned a doctorate, they understandably sought positions at colleges and universities where research received support and encouragement. "It is unfortunate," wrote Horsley in 1937, "that many of our scientifically trained young people leave the State to go where there are better opportunities and where the prestige of scientific research work is greater."[47] Three years later, George Palmer, professor of chemistry at the University of Alabama, echoed Horsley's sentiments: "If we train Southern men in scientific research, a certain proportion of them will remain in the South. Hence, the universities and colleges should take the lead in building up scientific research in the South. We should strive to have at least one institution in each of the Southern states granting the Ph.D. and M.D. degrees."[48] Palmer's dream would not be realized until the 1950s, though; until then the drain of talent from much of the South continued unabated, leaving a vacuum in the instigation of progressive research.

Finally, at a time when research was just emerging as an essential element of an academic scientist's career, a number of college and university science instructors chose either to ignore it or to make excuses for their lack of effort. Most often they pointed to the relative poverty of the South and the resultant lack of funds and equipment at their disposal. While this situation was not a figment of their collective imagination, more ambitious scientists refused to be so easily defeated. As the president of the Georgia academy, S. M. Christian, wrote to his colleagues in 1943: "We can subside into routine teaching and daily work, or we can launch research programs as significant as those anywhere. Let us never attribute our failures to freight rates or overwork, if the first enemy is inertia."[49]

Despite Christian's pleas and the efforts of many of the academies to encourage widespread competition for the available research grant money, most of the funds went to a small group of people, frequently those who were instrumental in the workings of the academies. These men, and occasionally women, represented the South's premier scientists, those with doctorates in their respective fields and positions at major colleges and universities. Their education had imbued them with the spirit of inquiry that translated itself into the goals of the academies. In general, they were the members best prepared to develop research projects worthy of funding. Miller indicated as much in a letter to Horsley in 1931, when he wrote that the "younger scholars could hardly expect to compete with more experienced ones, and thus give up on trying for the award."[50]

At the same time, relatively few of these more accomplished educators applied for the funds, either. Many of them, such as the Kenan Professors at the University of North Carolina and others who enjoyed some institutional research support, no doubt concluded that the small sums simply were not worth the paperwork involved. Consequently, the few who did apply for these awards usually received them. Obviously the research grant programs of the state academies did not operate as many people had hoped. Except in the case of Virginia, they did not attract additional funds, nor were they especially successful in encouraging research among a wide number of scientists. Thus they cannot be considered a significant factor in the professional lives of southern scientists.

Attracting the Interest of Industry

Many academies sought to attract the attention of industrialists who might be willing not only to join the organization but to support it with exhibits at the annual meeting and advertisements in the fledgling journals. Some southern industries maintained modest research facilities, and the state academies also hoped that scientists thus employed might join their ranks. With a broadened base of support, reasoned the academies' leaders, the organizations might hope to produce more creditable publications, enrich research funds, and enjoy more leverage in the state capitals as they fought for conservation and education issues.

Alabama scientists were most active in this regard. Of course in the 1880s and 1890s Eugene Allen Smith had set a tone of cooperation between academic scientists and industrialists as he shared his knowledge of Alabama's geology with the men bent on mining the rich coal and iron ore deposits in the northern section of the state. Just as the founders of the Alabama Industrial and Scientific Society of the 1890s had hoped to join the interests of academic scientists with the state's expanding mining industry, the Alabama Academy of Science did likewise. When George Fertig, a scientist employed by the Pittsburgh Testing Laboratory (located in Birmingham), became president of the Alabama academy in 1931, he immediately set out to expand the organization's membership by encouraging industrial scientists to join. He also persuaded the academy to inaugurate an industrial section, hoping not only to attract the interest of scientists outside academe but also to acquaint college professors with the particular problems and possibilities facing industry in the state and the South.[51]

Evidently Fertig failed to generate much interest in the academy among his industrial colleagues. Emmett B. Carmichael, a faculty member at the University of Alabama who had preceded Fertig as president of the academy, quipped a bit peevishly to Fertig's successor, J. F. Duggar, that Fertig had "never signed up a member [from the industrial community]."[52] In fact, total academy membership declined somewhat during the early 1930s from a high of 165 in 1930 to a low of 109 in 1936, before beginning to rise again. As with the Alabama Industrial and Scientific Society thirty years earlier, industrialists, whether managers or scientists, exhibited little sustained interest in the organization.

Scientists in Virginia also sensed the increasing force of industry in the South, knew that some companies already carried full-time scientists on their payrolls, and sought a closer relationship with these men of the nonacademic world. In 1930 the Virginia academy organized a committee "to study methods for establishing more complete understanding, better mutual relations, and greater cooperation between the Academy and the industries of the State."[53] Colonel Edwin Cox of Richmond succinctly pointed out "that if the mutual relations between the Academy and the industries of the State were more cordial and helpful there would be no difficulty in increasing the [academy's] income from that source."[54]

The Industrial Committee of the Virginia academy genuinely wanted to interest industrial scientists in the academy meetings, but it did not comprehend accurately the gulf that separated these persons from those who served as college professors. The committee assumed that industrial scientists were overwhelmed by the mass of technical papers presented at the meetings and would exhibit greater interest if more papers reviewed recent "notable progress" in various fields. The committee also noted that some academy members thought that papers presented by industrial scientists would be "potentially destructive of the pure scientific atmosphere of the Academy."[55] Although the committee offered a total of twelve recommendations designed to entice industrial scientists to join the organization, including symposia that would interest both industrialists and academicians, fewer technical papers, a separate industrial research award, and better publicity, the Virginia academy implemented few of these recommendations, and membership from the industrial community lagged.

The other southern academies made little specific effort to attract members from the industries of their respective states, and academic scientists continued to dominate the organizations. Of the 226 members of the North Carolina academy in 1935, fully 187 of them were affiliated with a college or university, 21 were government employees, usually with the state department of agriculture, 8 taught in high schools, and 10 listed no formal affiliation. The 1936 membership roster for the South Carolina academy indicates that of 234 members, some 52 percent were college pro-

[margin note: Few academies actively courted industry]

fessors, while an additional 19 percent were college students. The restrictive membership requirements of the Georgia academy until after World War II virtually precluded anyone but academic scientists from joining the organization.

In fact, most academic scientists failed to grasp the difference in outlook between themselves and their industrial counterparts. Much of the latter's work prior to World War II involved well-defined engineering and technical improvements directed and funded by their employing institutions. Obviously, industrial scientists did not share their academic colleagues' need for research funds. Nor did they require an outlet for the presentation of their research results, since continued employment depended more on successful implementation than on professional recognition. Thus the state academies, attuned to the concerns of the academic scientists who had founded and supported them, offered little of interest to the majority of industrial scientists.

Obviously, the state academies themselves did not operate as their founders had hoped. Concerning the three most pressing needs of southern scientists in the years preceding World War II, the state academies of science should receive high marks for providing the means for personal contact with one another, but they did not fare well in their attempts to publish creditable journals and stimulate research. Before categorizing these organizations as just another example of southern retardation, though, it should be noted that state academies in other regions of the nation were in many instances no more successful in meeting the professional needs of their members than were their southern counterparts. Very few state academies published a strong quarterly journal by 1940, and those that did usually received state financial aid. Three of the strongest academy publications, those of Illinois (first published in 1908), Indiana (1891), and Michigan (1899), contained complete papers beginning with the first volume, but even with state support they appeared only annually. The California Academy of Science published a commendable series of *Bulletins*, *Proceedings*, and *Occasional Papers*, but its handsome endowment fund put it in a league by itself. Most of the state academies struggled to publish anything at all, and the results usually appeared simply as bulletins, programs, or abstracts, offering complete papers rarely, as funds allowed.

Nor were state academies of science outside the South any more successful in providing their members with research funds. Every state academy that affiliated with the AAAS was eligible for its modest grant program, but few of these state organizations followed the lead of Virginia in enhancing these funds. The Indiana Academy of Science maintained a small endowment fund that in 1936 had assets totaling $6,200. The following year its research committee used $230 earned in interest to supplement the $100 received from the AAAS and awarded a total of five grants ranging from $50 to $75. In the same year, the Illinois academy offered three grants, all less than $100, provided entirely by AAAS money. Ten years later, these amounts had not increased appreciably. In 1948 the Illinois academy awarded $408 for research, and the Indiana academy offered $400 for that purpose.[56]

Although limited, this comparison does indicate that the southern state academies of science cannot be accused of disregarding the interests of their members while scientists elsewhere reaped great benefits from similar societies. The state academies in the South did emerge later than those in other regions of the nation, but this was more a function of slowly developing institutions of higher education than it was a belated realization on the part of southern scientists that they were being left behind. Once organized, the southern state academies of science developed similarly to organizations throughout the nation, and if they did not accomplish all of their goals as originally outlined, neither did many of the others.

Ironically, of course, because the state academies did not become the strong organizations envisioned by their founders, they could not attract the support necessary to develop as such. They were not, though, insignificant organizations. Many scientists received sustenance from the annual meetings at a time when they could not attend national ones; they had an opportunity to present the fruits of their research; they could at least keep up with what their colleagues around the nation were doing through the journal exchange program. Moreover, these academies gradually brought into their fold a number of persons teaching at small colleges, many with only bachelor's or master's degrees, whose education had not exposed them to the rigors of the new professionalism. While only a handful were likely to be inspired to attain the doctoral degree or pursue a research goal, many more than would have been the case otherwise experienced enough professional contact to keep them alert intellectually and to encourage them to pass along to their students the spirit of an inquiring mind. Such an impact is impossible to measure, but it was perhaps the most important single accomplishment of these organizations during their first two decades.

Southern Scientists   **6**
and the Ideal of Service

At least in part because of the activities of the state academies of
science, southern scientists by the 1930s had more opportunity
than ever before to feel a part of a professional community beyond
their individual campuses. As Nathan Reingold has noted, however, the concept of professionalism encompassed more than en-
forcing high standards of research and teaching and advancing
knowledge through publication. "The common definitions of a
profession," he states, "assume an applied component requiring a
service ideal."[1] Southern scientists did not have to look far to find
areas where they could be of service; conservation of the region's
natural resources and the quality of science education in southern
schools begged for attention. While all of the academies expressed
concern over these issues, the extent to which academies were able
to translate talk into action usually depended on the concerted
efforts of a few dedicated individuals.

Conservation of the Nation's Resources

With the rapid expansion of industry and the explosive population
growth in the United States following the Civil War, the con-
tentment of American naturalists with the beauty and abundance
of the nation's resources turned to alarm. Virgin timber was felled

with no thought of replacement; the lack of hunting regulations spelled the destruction of entire ecosystems; strip coal mines left ugly gashes on a once-pristine landscape. Conservationists turned to the government to halt this rape of the land, suggesting both permanent preserves and federal and state regulation that would oversee informed rather than wasteful use of the nation's resources. The reform atmosphere of the Progressive Era and conservation-minded Theodore Roosevelt in the White House promised hope.

Before Roosevelt left office in 1909, the federal government had set the precedent for regulation. The Reclamation Act, the Newlands Act, and the tremendous expansion of national forests, along with the controlled use of their resources, paved the way for future conservationists to demand even further government action. Roosevelt brought this broad issue to national attention when in May 1908 he called a conference of governors to lay before them the immediacy of the situation. One month later he authorized the formation of the National Conservation Commission, which by December had produced an inventory of the nation's resources that predicted dire shortages should preservation action be neglected. While many of the estimates concerning quantity of reserves and rate of use were grossly inaccurate, this report awakened much of the nation to the consequences of a plundering mentality.

Why was the nation willing to trust the pronouncements of such government commissions? By the early twentieth century Americans had become accustomed to the presence of "experts" in their society, men and women with specialized knowledge in a specific area, and were willing to cede a certain amount of authority based on this knowledge. Millions of industrial workers had encountered this philosophy when factory managers adopted Frederick W. Taylor's theories of scientific management. Reform legislation of the Roosevelt, Taft, and Wilson administrations depended heavily upon the advice of lawyers, social workers, and economists as well as upon public demand. Similarly, the general public increasingly regarded scientists as experts and accepted statements that began "Science tells us . . ." as the unquestionable truth. While such trust was often abused, as when the American advertising industry used pictures of men in white laboratory coats to sell all manner of products, it also offered scientists a golden opportunity to utilize their specialized knowledge for the public good and to gain a national forum. Hence pronouncements from the AAAS, the NAS, and other national professional organizations, as well as from individuals, usually captured a portion of the front page of the nation's newspapers. While activist scientists numbered only a small minority among their colleagues, they garnered considerable attention and often at least tacit support from fellow scientists on the issues that they adopted.[2]

During the 1920s and the 1930s, southern scientists found that the state academies of science provided a platform for them to address issues of specific local concern, with the environment frequently one of those issues. In part their interest was economic. Far from distraught over the growth of industries that fed off the region's timber and minerals, they saw these companies as contributing to the South's economic growth and asked only that proper techniques of extraction and, in the case of forests, replacement, be observed. However, southern scientists did express alarm when growth or neglect threatened such important and irreplaceable natural habitats as the Great Dismal Swamp, which spanned the Virginia–North Carolina border, and the wilderness areas along the eastern coastline. They also fought to preserve certain areas simply for their scenic beauty or for their potential benefit to scientific research. While the academies of Tennessee, North Carolina, and Virginia were most successful in implementing conservation programs, they were joined by others who voiced concern over the same issues.

In 1912, its first year of operation, the Tennessee academy petitioned the governor to form a conservation commission to study the issues of waterpower, forest conservation, and the ecological abuses of the mining industry. Later it encouraged the government to purchase land in the Great Smoky Mountains for a park.[3] Many papers read before the early meetings of the Tennessee academy, although not directly concerned with use or misuse of natural resources, emphasized the importance of a thorough knowledge of such diverse areas as the meadows of the Cumberland Plateau, the fisheries of the state, and phosphate and other mineral beds. In 1915 R. S. Maddox, the state forester and a member of the academy, made a presentation to his colleagues entitled "Preservation of Our Forests." Although the text of his remarks has been lost, it was probably similar to the comments of his counterpart in North Carolina, Joseph Hyde Pratt, who emphasized not only the significance of forestry resources and the need to conserve them through selective cutting and replanting but also the importance of the forest for dependent animal life.[4] After 1925 the Tennessee academy devoted most of its attention to Reelfoot Lake, as discussed in the previous chapter.

Although the North Carolina Academy of Science sponsored no program equal to the Tennessee academy's biological research station, it nonetheless expressed a considerable concern for the South's natural resources. Joseph Hyde Pratt, head of the North Carolina Geological Survey, led the academy's conservation efforts in the early years through a number of articles in the *Journal of the Elisha Mitchell Scientific Society.* In "The Southern Appalachian Forest Reserve," for example, Pratt outlined the need for governmental control over timber operations, while in an address entitled "The Occurrence and Utilization of Certain Mineral Resources of the Southern States," he urged wise use of these potentially valuable elements. To these pleas he added a paper entitled "The Conservation and Utilization of Our Natural Resources," his summary of Roosevelt's May 1908 Conference of Governors, to which he had been invited.[5]

The North Carolina academy took no immediate action on any specific issue, and in 1917, when Pratt joined the United States Army, the catalyst disappeared. Upon his return, however, he was elected vice-president of the academy, and at a meeting of the executive committee in 1921, he recommended the formation of a committee on preservation of natural resources. The membership approved the recommendation, but the committee did not evolve as Pratt had hoped. Two years later it did suggest that the academy consider publishing a series of nature study pamphlets and that it might cooperate with the Virginia Academy of Science in "securing the conservation of a selected part of the Dismal Swamp."[6]

Again, nothing immediate came of either of these two recommendations, and the conservation committee lapsed until 1937.

Despite the lack of specific action, the academy continued to express a concern for conservation principles, leading to a number of resolutions during the 1920s and 1930s. In 1928 the members voted to support the McNary-McSweeney Forest Research Bill and forwarded copies of their resolution to North Carolina congressmen and other governmental officials. In 1935 the members formally protested what they viewed as the "apparently ruthless and unnecessary destruction of thousands of trees and shrubs" by overzealous Civilian Conservation Corps workers. Again the academy brought its concern to the attention of government officials, including President Franklin D. Roosevelt and the governor of North Carolina.[7]

In 1937 a reinvigorated Conservation Committee, under the leadership of J. S. Holmes, a forester with the North Carolina Department of Conservation and Development, issued a four-page report on the conflict between the economic use and the limited availability of natural resources. Pointing out that two-year planning commissions could not resolve a Pandora's box so fraught with serious implications for the future, the committee called for some serious long-range planning. It noted that locks had been built in the rivers with no thought for the shad that inhabit the same waters, and it declared that too many laws seemed to favor hunters and their "misplaced notion" of conservation. Above all, the committee concluded, "the great task before us is the conversion of a wasteful and indifferent public, to one recognizing its duties to the future, before it is too late."[8]

Throughout the next year the committee analyzed various natural areas across the state and listed a number of those worthy of preservation. Among its special concerns was Black Mountain, which the committee asked the United States Forest Service to dedicate as a natural area in perpetuity. It also requested the government to purchase the Ravenel Forest and Richardson Woods, a "primeval forest" near Highlands, for the same purpose. In its report of 1938 the committee indicated success with the Black Mountain project, but it noted that the government would not purchase the Ravenel Forest. Upon the owner's death in 1940, the heirs asked one hundred dollars per acre for the land, a price the government was unwilling to pay. Eventually much of the area succumbed to a nearby lumbering operation.[9]

The Conservation Committee continued thereafter to act as a natural resources watchdog, reporting potential problems to the academy membership and asking for formal recommendations in order to call such problems to the attention of government officials and the general public. On the whole, the committee enjoyed success in its efforts. In 1943, for example, it announced that the Dismal Swamp, hungrily eyed by land developers, had come under federal protection. Other conservation crusades involved protection for the wood duck, whose multicolored plumage was much in demand by the millinery industry, and sanction for wild Venus's-flytrap plants, which evidently were being spirited out of the state illegally and sold as novelties. One of the committee's most satisfying victories came much later, in 1953–54, when it swung into operation to save Crabtree Creek Park, located between Raleigh and Durham, from becoming a United States Air Force base. Eventually the government built Seymour Johnson Air Force Base farther east, near Goldsboro.[10] Probably the academy claimed too much credit for this feat, because it was by no means the only group opposed to the initial site choice. Then too, the government doubtless had selected several suitable locations and, in the face of opposition, chose a path that promised less resistance.

The Virginia academy's concern for conservation began in 1925, when the Ecological Society of America asked the organization to help defeat a bill pending in the United States Congress that would lessen the control of the Forest Service over grazing rights on government land. Although the academy took no formal action, it suggested that individual members write their congressional representatives concerning the matter. One year later the academy created a permanent committee to oversee the protection of Virginia's natural areas and thus initiated what became a long history of concern for irreplaceable resources.[11]

Operating much like the committee of the North Carolina academy, the Virginians identified locations throughout the state worthy of preservation and simultaneously advocated a coherent, long-range state conservation program. Watching the development of the state for possible infringements on their targeted areas, the academy in 1929 resolved to fight the construction of a dam by the Virginia Public Service Company at Goshen Pass, a gorge in the Appalachian mountains. It again swung into action by petitioning state legislators when real estate developers suggested that the Great Dismal Swamp be filled in and thereby rendered "useful." Although the academy's contributions to the resolution of these issues is difficult to measure, Goshen Pass remains today in its natural state, and the Great Dismal Swamp has not yet become a part of the concrete jungle.[12] In 1930 the Virginia academy joined with various garden clubs of the state, the Izaak Walton League, and the state Commission on Conservation and Development to supply the state's administrative officials with an advisory council relative to state parks and forests.[13] Generally speaking, the Conservation Committee remained alert, and when an immediate problem arose, it took appropriate action.

The Georgia and South Carolina academies adopted a more modest approach to conservation. Neither organization formed a conservation committee, but the Georgia academy did pass resolutions from time to time when matters of conservation were brought to its attention. At its initial meeting, the academy declined to act on J. M. Reade's suggestion that it advocate preservation of the Okefenokee Swamp, but in 1924 it did go on record in favor of preserving Glacier Bay, Alaska. Later such recommendations involved the Florida Everglades and Stone Mountain, Georgia, and in 1963 the Georgia scientists supported a movement to create a state board for the preservation of natural areas.[14] The Georgia academy's lack of positive action no doubt stemmed from its somewhat unusual organizational structure. With a limited membership of fifty persons and stringent qualifications for admission, it operated more as a science fellows club than as a service organization until after World War II. The South Carolina academy, meanwhile, was struggling just to survive.

Alabama's academy tied most of its concern for conservation directly to the promotion of industrialization in the state. While numerous members of the academy actively endorsed the development of the state's iron and coal resources, they did so on the basis of a policy of wise use rather than reckless extraction. Although they did not completely understand all of the economic ramifications of the type of industry they promoted, they never intended for industry to offer jobs to Alabamians in exchange for exploitation of the environment. Moreover, they correctly foresaw that in order for the state to realize the full potential of industrial development, it would have to include research as well as new jobs. Their story more appropriately is told in the next chapter.

A Boost to Secondary Education

While the southern state academies of science varied in their approach to and support of the conservation of natural resources, they all agreed on one matter of vital importance to the South—science education begged for improvement. Although scientists in Georgia and South Carolina were once again less vocal on this issue than their other southern colleagues, they too evinced concern for the region's schools, particularly the nature of science education offered in secondary schools. Since the majority of the region's scientists taught in colleges and universities, they witnessed firsthand the results of inadequately prepared students from the South's high schools. Before the onset of World War II, most of the academies had initiated efforts to improve the quality of science education in the public schools. Some of them also adopted programs aimed directly at the students themselves, including the support of school science clubs, the formation of junior academies of science, and a visiting scientist lecture program. These student-oriented activities did not mature fully until after World War II, but such programs were a product of the realization during the 1920s and 1930s that the South's future rested in the hands of its youth.

The North Carolina academy was among the most active in seeking to improve the quality of secondary science education. Beginning with Trinity College physics professor C. W. Edwards's remarks in 1909 that bemoaned the inadequate academic preparation of college freshmen and led to the formation of the academy's

Committee on Science Teaching and its various reports published in the *North Carolina High School Bulletin,* the North Carolina academy mounted a three-pronged approach to improving science education. First, it encouraged high school science teachers to join the academy, and it promised that individual members of the academy would meet with science teachers at the annual gatherings of the North Carolina State Teachers Assembly. Second, it called for establishment of a sound working relationship between the academy and the state board of education in order to facilitate the free flow of ideas and the adoption of a number of proffered suggestions. Third, the academy initiated several programs through which it established direct contact with high school students and sought to foster their scientific curiosity.

In 1917 Bert Cunningham, a professor of biology at Trinity College and a former high school teacher, suggested that the academy try to attract more high school teachers into its fold, perhaps by dedicating a special section specifically to their interests and concerns. Although Cunningham's suggestion produced no immediate action, the academy sent a letter to all high school science teachers in the state in 1923, asking them to attend the annual meeting of the state Teachers Assembly and to form a separate section for science teachers within that organization. C. E. Brooks of the North Carolina Department of Education supported this idea and encouraged teachers to participate in the program. During succeeding years various North Carolina scientists met with the high school teachers and presented a summary of the latest developments in their fields of interest. North Carolina high school science teachers thus gained a forum for discussing their concerns and problems and created as well a united front for making specific requests to the state board of education.[15]

In addition, Brooks agreed to organize a committee composed of ten professors from six colleges under the auspices of the Department of Education to assess the specific science courses taught in the state's high schools, including availability of equipment and quality of instruction. The members of this committee received authorization to discuss with teachers and administrators specific problems and to suggest possible solutions. Additionally, the

board of education charged them with the responsibility for noti-
fying colleges of the need for courses designed to train science
teachers, both in content and method, and to encourage colleges
to place these courses in various science departments rather than in
the education department.[16]

Taking its duties seriously, the committee soon presented to the
North Carolina academy a number of recommendations to be
passed on to the Board of Education, all of which received hearty
endorsement. The committee supported Brooks's expressed desire
that the state require at least one science course for high school
graduation and that each school be equipped with at least a minimum amount of laboratory equipment. It also recommended that
the state specify a minimum amount of scientific training for
teachers in the subjects they expected to teach and proposed a differential pay scale in which better teachers (meaning those with
specific scientific training) would receive a higher salary. To encourage prospective treachers to take additional science courses,
the committee suggested that college students planning to teach be
allowed to substitute two advanced science courses for one education course. Finally, it called for continuation of school visitations
and recommended that the state hire a full-time supervisor of science instruction.[17]

While the working relationship between the Board of Education
and the North Carolina Academy of Science continued to be productive for many years, financial realities and the ponderous
nature of bureaucratic operations dictated that many of the academy's specific recommendations would fall by the wayside. The
organization was most successful when reaching out directly to
teachers through the Teachers Assembly and passing along to
them the most recent developments in botany, zoology, geology,
chemistry, and physics. While the state did not immediately adopt
the academy's recommendations concerning minimum science requirements for high school graduation, necessary apparatus, and
college entrance requirements, these recommendations did serve
as a foundation for future action.

In addition, the North Carolina academy also sought to stimulate an interest in scientific inquiry among the high school students

themselves. In 1927 the organization inaugurated an essay competition for high school science students. Each school in the state could submit up to three entries to a panel of judges appointed by the academy. The prize, a loving cup, was presented to the winner each spring at the high school graduation exercises by a member of the academy.[18]

The contest enjoyed limited success at best. The greatest number of entries in any one year was only forty-six (in 1933) and represented just twenty-six high schools. Until this time the number of entries had steadily increased, and the membership, pleased with the response, voted to change the prize from a loving cup to books. Even so, they were less pleased with the quality of the essays. In 1932 the committee recommended that "papers be urged to avoid reproduction of articles and book passages in form similar to their original, and the holding of direct quotations to a minimum."[19] Five years later, when the number of entries declined to twenty-one papers, the committee tried to increase participation by communicating directly with the science teachers concerning the contest rather than going through administrative channels. The results were not encouraging. Not even a cash award of twenty dollars offered by the North Carolina Forestry Association for a separate essay contest generated much interest.[20] When the academy resumed normal operation after World War II, it abandoned the essay contests in favor of other efforts to stimulate scientific interest among high school students.

The members of the Tennessee Academy of Science likewise expressed concern over the quality of scientific education in the public schools. In 1926 they rose up in arms over the passage by the state legislature of the now infamous Butler Bill, which effectively banned the teaching of the theory of evolution in the state's biology classes and led to the trial of John Thomas Scopes in Dayton later that year. Terming the law "an unfortunate limitation of the intellectual freedom of teachers of science in our public schools," the academy insisted "that it marks a backward step in our educational program; that it takes away important privileges heretofore available to students, especially those in our higher institutions."[21] The narrow conception of scientific inquiry as reflected in this legislation went against the grain of everything for which these scientists stood. Battling to encourage high school and even college teachers to keep up with the latest developments in their fields and pass them along to their students, scientists were appalled over the stubborn insistence on adhering to a literal interpretation of the Book of Genesis. A resolution adopted unanimously by the academy called for repeal of the bill during the next legislative session. Once the sun had set on the Scopes trial, though, the matter was largely forgotten, and the bill remained law until quietly repealed in 1967, just one year before the Supreme Court, in *Epperson v. Arkansas*, ruled such legislation unconstitutional.[22]

The Tennessee academy was not alone in abhorring the resumption of the battle over the theory of evolution. Nonetheless, southern scientists proceeded cautiously in their opposition. In 1923 George O. Ferguson, professor of psychology and education at the University of Virginia, suggested to Ivey F. Lewis, president of the Virginia academy, that the organization should prepare some sort of statement in case "Mr. Bryan comes this way." In April 1924 Lewis polled the members of the executive committee concerning Ferguson's suggestion. E. C. L. Miller, the academy's secretary, responded: "It is hard for me to see the necessity for a statement concerning evolution because with me it is as commonplace and fundamental an assumption as that the sun will rise tomorrow. However, I suppose that there are many people who are 'gun shy' on evolution. It is a question whether any statement we could make would have any effect on them." Counseling caution, Miller suggested that perhaps the Virginia academy could issue a statement that the theory of evolution was regarded as a working hypothesis, that it should not be termed a "doctrine," and that the word "believe" should not be used in connection with theory because such terms smacked too much of religion.[23] Henry Smith, president of Washington and Lee University, also advised caution. While he supported the idea of a "very carefully worded resolution," he nonetheless pointed out that "in such an ultra-conservative state as Virginia, it might excite the extreme fundamentalists, who I fear are quite numerous in the Old Dominion, to greater alarm and more violent efforts at repression than ever."[24]

In his presidential address to the Virginia academy in 1924, Lewis spoke directly to the fundamentalists when he quipped that since stoning blasphemers was no longer considered acceptable Christian behavior, perhaps other elements of Christianity could change with the times as well. If Lewis hoped to arouse more of his colleagues to action, however, he was sorely disappointed, for although many of them privately expressed dismay over the situation, the academy passed no resolution relating the theory of evolution to freedom of academic inquiry.[25]

Although evolution did not become a major public issue in Virginia, it certainly did in North Carolina. Before the dust had settled, antievolutionary crusaders had tried to discredit both William Louis Poteat, president of Wake Forest College, professor of biology, and former president of the North Carolina Academy of Science, and Harry Woodburn Chase, president of the University of North Carolina. Both men had taken a stand in favor of freedom of thought and against the passage of a bill similar to that approved by the Tennessee legislature. At its annual meeting in 1926, during the height of the controversy, the scientists of the North Carolina academy unanimously adopted the following resolution:

The North Carolina Academy of Science desires to reiterate that if the present rate of progress and enlightenment in the state of North Carolina is to be maintained and advanced, it is absolutely and unqualifiedly necessary that all those hypotheses, theories, laws and facts which constitute the legitimate content of any field of study, may be dealt with at any time by any teachers.

The Academy goes on record as endorsing most emphatically the stand of Dr. H. W. Chase and Dr. W. L. Poteat on the freedom of thought and teaching.

Although the impact of the academy's resolution amid the plethora of petitioners is difficult to measure, the North Carolina legislature by a narrow margin defeated a bill banning the teaching of evolution.[26]

After expressing outrage over the public effort to censure the theory of evolution, the Tennessee academy thereafter confined its concern with science education to papers that addressed various aspects of science teaching. Although it continued discussing text-book content, methods, and teacher training, the academy did not specifically solicit high school teachers for membership, nor did it organize a separate section for educators. The members did discuss supporting high school science clubs as early as 1930, but they took no positive action until 1940, when they invited members of the Texas Academy of Science to address their annual meeting concerning the junior academy of science in that state.[27]

The Tennessee academy could have invited speakers from a state much closer to home to discuss the formation of a junior academy of science. By 1940 Alabama's junior academy had been organized for seven years and met annually in conjunction with the senior academy. In 1932 Emmett B. Carmichael, a professor at the University of Alabama School of Medicine, suggested a junior academy to his colleagues after hearing Howard Enders of the Illinois Academy of Science give a glowing report of that academy's success with the junior academy, organized in 1919. Upon Carmichael's recommendation, the membership in 1932 formed a committee to study the matter.[28]

Carmichael, John R. Sampey, and Russell S. Poor began by requesting information from those academies that had organized junior academies, particularly Iowa and Illinois. As they made contact with state high schools, they soon discovered a number of local science clubs through which they could work and began to funnel their accumulated literature on junior academies into this network. Sampey and his colleagues, expecting an enthusiastic response, dishearteningly noted in December that they had received only lukewarm reception of the idea.[29] In January 1933 Sampey, Poor, and John Xan personally visited the high schools in the Birmingham area to meet with teachers and students and discuss the idea further. They pushed for a junior meeting in March in conjunction with the senior academy and encouraged each club to send representatives, speakers, and exhibits. Incentives included the possibility of prizes for outstanding papers and exhibits.[30]

The initial meeting of the Alabama Junior Academy of Science convened on March 11, 1933, in Birmingham. Ten high schools, five of them outside the Birmingham area but located in the northern half of the state, sent representatives. Given the financial depression at the time and the continued difficulty and expense of long-distance travel, organizers expressed pleasure over the attendance. Emmett Carmichael spoke to the young members about the junior academies in other states, and two other members of the senior academy gave brief talks on scientific subjects. In addition, a number of students presented papers, many of them brought exhibits, and the assembled membership elected officers and approved a constitution.[31]

The junior academy met jointly with the senior academy throughout the 1930s, suspending their sessions only during World War II. The senior academy maintained an advisory committee composed of two counselors appointed annually by the president and joined in 1937 by James L. Kassner, appointed permanent counselor to ensure continuity. Counting on high school teachers to encourage their students to participate in the activities, the senior academy did offer small cash prizes, usually five or ten dollars (depending on the status of the treasury), to the three best student papers and to the best exhibit in each of the fields of chemistry, biology, and physics. The programs of the junior academy usually included one or two members of the senior academy, who discussed matters of interest to the students. An active organization prior to World War II, the Alabama junior academy recovered quickly from the disruption of the 1940s and soon became the most vibrant and imitated organization of its kind in the South.[32]

The Virginia academy also expressed an interest in improving science education. As early as 1925 the academy attempted to increase its membership by appealing to high school teachers. E. C. L. Miller attended the science section of the Virginia State Teachers Association that year and extended membership to all of those present. He reported somewhat disappointedly that no one seemed very enthusiastic about the opportunity.[33] From then until after World War II, the Virginia academy, while always willing to welcome high school science teachers into its membership, exerted no great effort to attract them, nor did it continue to send representatives to the state teachers' assembly.

Likewise, the 1930 suggestion of forming a junior academy of science produced considerable discussion but no concrete action. The academy did appoint a standing committee on science in the public schools "to consider the part that training in sciences should play in the process of education, the time and attention in school programs that its importance justifies and, from time to time as need and opportunity appear, to exert all proper influence to improve the standing of the sciences in the schools of the state."[34] Unfortunately, this committee never established a working relationship with the state board of education, and so its suggestions remained on the drawing board.

The academy did not, however, totally abandon its interest in the young people of the state. The Committee on Science in the Public Schools continued, and by 1938, largely through individual efforts of the members rather than concerted academy action, it reported that it had established contact with some high school science clubs and hoped to be able to help them in some manner. In 1940 the academy, aware of the growth of junior academies throughout the nation, decided to inaugurate one in its own state. The newly formed junior academy first met in 1941 and, like that in Alabama, became one of the most important aspects of academy work in the post–World War II era.[35]

The attempts of southern state academies of science to improve science education and to foster a spirit of conservation met with limited success at best. Just as the relatively young organizations often encountered difficulty in meeting the professional needs of their members, so too were their efforts to embrace social responsibility sometimes thwarted. Although most of the academies duly organized committees on conservation and education, any action on their part, other than producing formal resolutions, required both time and money. The individual members had little of the former to spare, and the academies generally operated with very small treasuries.

Why, when the academies constantly complained that they lacked sufficient funds to expand their programs, did they not seek more aggressively either state or private funds to aid them in these endeavors? The answer is not simple. In the first place, petitioning legislatures and filling out grant applications, particularly during a time of financial depression, must have seemed to many academy members as an exercise in futility. The North Carolina academy at one time voted to seek state funds, but its executive committee gave up the matter without a whimper, "owing to the financial stringency existing in the state at the time."[36] Compounding the problem was their press for time, as they faced a full complement of students and laboratory instruction as well. Then, too, many members (and potential members) were new to the spirit of professionalism. Again, it was a matter of time before they would appreciate fully the stated goals of the academies. Finally, employment, particularly in the smaller schools in the South, did not yet depend as completely as it would later on professional achievements and activities. In short, the academies could not emerge as fully mature organizations overnight, and the development process seemed to be more lengthy and involved than many had supposed.

During the 1920s and the 1930s, then, the state academies of science grew slowly and experienced difficulty implementing both their professional and service goals. By the late 1930s, just as many of them were reaching the level of maturity necessary to sustain such activities as the publication of a journal and the support of junior academies of science, the demands of war forced them to turn their attention to other matters. The effects of the war on the world, the nation, and the South changed drastically and forever the environment that had fostered these state academies. While all of the academies emerged intact following the war, the profession that they served was so altered that new emphases had to be found if they were not to wither and die.

7

A Dream Realized?
Advent of Research Facilities
in the South

In retrospect, World War II marked a significant turning point for the scientific profession in America, and most especially for southern scientists. Wartime mobilization not only revived a depressed national economy but fused two forces grown quite powerful during the twentieth century—the United States government and scientific expertise. At first centered on the military-industrial complex, this union broadened during the subsequent decades to include government-sponsored research by private, nonmilitary corporations and government financial assistance for research conducted in institutions of higher education. Southern scientists realized full well that Dixie could profit enormously from this turn of events if only the United States government and national corporations could be convinced of the advantages of investing in the diversification of the former cotton kingdom. Armed with hope and good intentions, a handful of southern scientists set out to mobilize their colleagues and join forces with other southern boosters who sought economic revitalization through industrial development.

The scientists, unlike agents of the various southern state governments, chose to emphasize the potential of the region for research and development facilities rather than low taxes and a cheap labor supply.[1] By the 1940s these academicians and a few other farsighted individuals realized that labor-intensive, low-investment industries would not produce the self-sustaining growth pattern they coveted. Their promotional vehicle designed to attract governmental and industrial attention, the Southern Association for Science and Industry (SASI), was spawned by the maturing state academies of science. Hailed by its organizers in Alabama as the opportunity for southern scientists collectively to realize their much-cherished goal of research support, SASI soon mushroomed beyond their control into an organization that by 1945 very few southern academic scientists were willing to support. By this time university budgets had increased exponentially, graduate and research programs had secured funding, private research firms had opened, and multifaceted industry began moving south. The corner had been turned. But such rapid progress was in the unforeseeable future in 1940, when one Alabama scientist decided to take action.

Industrialization: The Hope for Economic Regeneration

Well before 1940 the members of the Tennessee academy expressed an interest in the potential benefits of industrial development. In 1912 the academy exhorted the lawmakers of the state to form a conservation committee that, among its other duties, would be the responsibility for enticing industrial firms into the state.[2] During subsequent years academy members presented a number of papers detailing Tennessee's resources and suggesting how they might be utilized. Following the creation of the Tennessee Valley Authority in 1933, the attention of the academy naturally centered on its implications for the state's development. A symposium on the TVA at that year's annual meeting attracted over one hundred persons to hear about the plans for power development, transportation improvement, agricultural rehabilitation, forest reclamation, flood control, industrial development, and social betterment.[3]

Aware that the vision for the area's development would take many decades to realize, A. E. Parkins, professor of economic geography at George Peabody College, nonetheless outlined for his fellow members of the Tennessee academy the possibilities of this dream in glowing detail. He remarked that "the creation of pros-

perous agricultural and industrial communities, peopled by con-
tented, healthy, educated, law-abiding, progressive citizens,"
could only evolve hand-in-hand with the harnessing of water, the
mining of coal, and the utilization of the region's other abundant
resources, including timber, limestone, marble, and minerals for
fertilizer production.[4] Similar to other regional planners, Parkins
portrayed the future of the Tennessee Valley landscape as "dominated by small industrial centers each with well-kept factories,
neat clean homes, schools, churches, and community houses; and
surrounded by garden, poultry, fruit, and dairy farms on which
the majority of the factory workers would live."[5]

With these remarks, Parkins displayed no more foresight in his
advocacy of industrial development for the South than did
boosters throughout the region who sought to attract industry to
their states. As historian James C. Cobb has noted, southern governors and local officials alike, in their search for economic
progress through industrial development, promised prospective
industries low taxes, a cheap labor supply, and abundant and
sometimes free utilities if they would locate their plants in the
South. This policy did little to attract high-technology industry to
the region, nor did it encourage the sort of development that
would become self-expanding. In short, concludes Cobb, it only
sustained the intricately interwoven fabric of southern tradition
that included white supremacy with domination by the members
of the upper class and minimal governmental interference concerning such matters as labor relations and industrial taxation.[6]

Nonetheless, southern scientists, troubled by the lack of institutional commitment to research, knew that without regional economic growth, even the bequests of philanthropists could not fill
the void. While many of the southern state universities had
accepted some responsibility for serving their own communities
through facilities such as agricultural extension programs, their
commitment to other research grew very slowly prior to 1945.
Thus many scientists joined other southern boosters to call for
industrial development that could offset the region's colonial economy.

As early as 1932 George Fertig, a Birmingham industrialist who was then president of the Alabama academy, did his best to spark this concern into action. Speaking to the academy on March 11, 1932, he blamed much of the South's economic woes on inadequate support for scientific investigation in the region's colleges and universities. "Whatever be the reasons for the slowness in the development of scientific departments in educational institutions," he said, "they had a telling effect upon the rapidity with which science was applied to the advancement of industry." Fortunately, according to Fertig, this situation had begun to change. Although he did not indicate the source of his information, he claimed that "three hundred and seventy-six researches are under way at thirty-one of the larger colleges [in the South]." On the other hand, he lamented, forty-six of the smaller schools reported a total of only thirty-one research programs, and an additional thirty-five schools reported no research at all.[7]

With the South's temperate climate, abundant rainfall, and "many raw materials of industry and its excellent facilities for all varieties of transportation of industrial products to the wide world," scientists, Fertig insisted, should turn their attention to research that would encourage industrial development and thus aid the region economically. Citing several southern research programs involving rayon, paper, cottonseed, fertilizers, naval stores, and steel and iron production, Fertig concluded, "May the splendid accomplishments of the last fifty years serve as an ever accelerating stimulus to finer academic and industrial attainments to the end that in all the fields of science she [the South] may currently come to contribute her proportionate share of fruits of scientific research."[8]

Russell S. Poor, a professor at Birmingham-Southern College, stated the situation even more succinctly. In his presidential address before the Alabama academy in 1935, Poor spoke on the South's position in the mineral industry. He remarked that "what the South needs now is the concentrated efforts of her trained sons. We now spend one billion dollars annually on food and feed from other sections." Poor concluded that research involving the South's abundant mineral supply would benefit the region's agricultural production and thus lead to a more balanced economy.[9]

The Alabama academy's concern with southern industrial development reached a peak when George D. Palmer, a chemistry professor at the University of Alabama, addressed his colleagues on March 29, 1940, taking as his theme "Scientific Research, the Hope of the South." Seizing upon President Franklin D. Roosevelt's identification of the South as the nation's number one economic problem, Palmer pointed out that "the present status of the United States as the leading nation is due primarily to the *unbeatable combination of business and scientific research,* backed by our great resources." The South, he averred, was underutilizing its resources because of a lack of research. To him the answer seemed simple enough. "We should develop our own scientific research organizations for work on Southern resources."[10]

"We often hear," continued Palmer, "of the absentee ownership of industries of the South, but we never hear about the absentee research done for these same industries." Although recent actions of the federal government, such as the establishment of the TVA, had contributed greatly to the stimulation of southern scientific research, the region could not depend on outside resources forever. Southern businessmen, Palmer insisted, should be awakened to the economic potential of research and encouraged to invest (not spend) their earnings in both industrial and academic research programs.[11]

Palmer, a native southerner, adopted this challenge to his colleagues as a personal crusade. In this same address he outlined a plan of action designed to gain the attention (and investment dollars) of industry nationwide. Realizing that no corporation would invest in an unknown quantity, Palmer envisioned the first goal of southern scientists to be that of advertising themselves. They had to promote their talent not only to the national scientific community, as had long been their goal, but also to American corporations, which increasingly relied on research and development for their future profits. What agency could better implement this plan than the state academies of science, he queried, with their various standing committees and journals, especially if the latter could be broadened to contain complete articles and could be published with some degree of regularity?

One can well imagine Palmer's audience on that March day in 1940. So what else is new? they must have asked one another. Expand the journals, former academy presidents had cried. Where's the money? came the reply from many a harried editor who had probably just finished explaining to the printer that the bill would be paid "soon." Formulate a research project, admonished the fortunate few who held positions at well-endowed institutions. Once again came the familiar replies: Where's the time? the equipment? the money?

Palmer remained undaunted. The product of a then small southern college (Clemson, in South Carolina), a veteran of teaching at another one (Guilford, in North Carolina), and currently a faculty member at a state university without a doctoral program in the sciences, Palmer paused only for breath. He suggested that in addition to using academy journals as promotional tools, southern scientists should combine their efforts into a regionwide organization to search out financial support for research. Linking his own aspirations with those expressed by the Southern Governors' Conference earlier that year when they inaugurated a "ten year program of economic and cultural enrichment," Palmer dreamed that by 1950 "we shall be able to report much progress in scientific research."[12]

Southern Association for Science and Industry

Palmer's suggestion for a federation of southern scientists was not new. In 1934 J. McKeen Cattell, owner and editor of *Science*, had suggested to the AAAS that it sanction local branches throughout the nation in order to stimulate an interest in the national association.[13] George Boyd, a University of Georgia zoology professor and member of the Georgia Academy of Science, had taken this idea to heart. He invited representatives from various southern states academies to meet together in Atlanta on March 23, 1935, to discuss the matter. Although six academies sent delegates, few of them expressed much enthusiasm for such an organization. J. L. Brakefield of the Alabama academy pledged his support, but the representatives from Virginia and North Carolina relayed to their own academies doubts that such an organization would prove to be of much benefit to them. The North Carolina representative did not make his reasons explicit, but Leonidas R. Littleton of the strong Virginia academy did. "It is my own personal opinion," said Littleton, "that State academies should be encouraged to develop, and where weak, several might combine." In short, state academies should consider such a federation only if they were not large enough to serve their members in a manner they deemed necessary.[14]

Although Boyd's efforts had come to naught, George Palmer was not to be discouraged. He wrote to Boyd just a week before his address to the Alabama academy, "As I see it, the way to start anything is to start it, no matter how small the undertaking may be in the beginning." His powers of persuasion held sway over his academy audience, for the executive committee that year agreed to host delegates from throughout the South who expressed an interest in organizing such a regional society at the academy's annual convention in Mobile in March 1941.[15]

In addition to academy members, Palmer's appeal struck a responsive chord with several representatives of the media. John Temple Graves, a journalist for the *Birmingham Age-Herald*, devoted an entire column to his enthusiasm for Palmer's ideas. Over the next several years the popular columnist would be one of Palmer's strongest advocates, lending his support through his literary talents to the concept of economic advancement through scientific research. "I have never known him [Graves] to give that much space to anyone before unless he was running for President of the United States," wrote Harry Jennings, a Birmingham lawyer. Not to be outdone, the *Birmingham News* also headlined Palmer's remarks, proclaiming that "Academy President Urges Dixie 'Wake Up' to Its Opportunities." The *Birmingham Post* quoted Palmer extensively as he provided numerous examples of outstanding scientific research then underway in the South, despite a woeful lack of funds and facilities. Columnist E. D. McCluskey concluded his remarks by reiterating Palmer's challenge to southern state legislatures to "appropriate additional money to the universities and other state colleges for scientific research."[16]

Palmer received national attention and generated considerable interest when his address appeared as the lead article in the July 5, 1940, issue of *Science*, the official organ of the American Association for the Advancement of Science. "I have been astounded," wrote Palmer, "by the number of people in the South and from other parts of the country who have been interested in my article." E. C. L. Miller, secretary of the Virginia Academy of Science, requested five hundred reprints of the article. Others congratulated Palmer and offered their assistance in implementing his suggestion that southern scientists organize themselves across state lines. In a letter to his friend and colleague C. M. Farmer at Troy State Teachers' College (Troy, Alabama), Palmer noted that "I am snowed under with correspondence" relative to the proposed organization.[17]

Wortley F. Rudd, president of the Virginia Academy of Science and a colleague of Miller's at the Medical College of Virginia, was among the first to inquire for more details from Palmer. "The question of research in the South," responded Palmer, "is one which concerns all Southern states. I have been intensely interested in expanding and helping the Alabama Academy of Science during the past year. Among many other things, the Alabama Academy of Science agreed to sponsor a meeting of delegates from the various state academies and other scientific organizations from the Southern states to organize a Southeastern Scientific Society." After outlining his plan for a meeting of such delegates during the following year, Palmer concluded: "I sincerely believe that the state academies are doing much for the Southern states, and that a Southeastern Scientific Society would greatly help us to know each other and our problems. . . . The more we can get people in the South thinking about Science, the better off Science will be in the South."[18]

Despite numerous pledges of support, Palmer had his work cut out for him. Between March 1940 and March 1941 he corresponded with officers of the various academies of science as well as with his Alabama colleagues concerning the type of program that should be offered at the meeting. Stewart Lloyd, dean of the School of Chemistry, Metallurgy, and Ceramics at the University of Alabama, suggested that Palmer invite Alabama's governor, Frank M. Dixon, to deliver the keynote address. Palmer wrote to Ernest V. Jones of Birmingham-Southern College that Governor Dixon "has been our chief trouble in the past. We as scientists have not played ball with the business interests and I think he is certainly right." He added, "I believe the Academy will be strengthened by joining its forces with the business men of Alabama. I am going to tell them so at the state chamber of commerce next week. Also, I am going to ask that the Ala. state Chamber of Commerce join hands with the Ala. Academy of Science in furthering science in the state."[19] The scientists were about to field a team.

Palmer did indeed speak to the state chamber of commerce in March 1941, at the invitation of Thomas W. Martin, president of Alabama Power Company. Presenting essentially the same message that he had published in *Science* the previous year, the eloquent Palmer appealed for the support of the state's leading business interests for his proposed organization and once again gained the attention of north Alabama's newspapers. Speaking in language that industrialists could well understand, Palmer insisted that "the present status of the United States as the leading industrial nation is due chiefly to the unbeatable combination of business and scientific research. . . . The concentration of research in the North," he continued, "amounts to a scientific research differential which, in the long run, is more important than our freight differential."[20]

In addition to industrialists, Palmer hoped to attract interest from the state academies of science; the regional associations of biologists, physicists, mathematicians, and meteorologists; medical societies; chapters of Sigma Xi (an honorary scientific fraternity); state and federal research laboratories; and teachers' organizations. In his correspondence Palmer cited a plethora of possible benefits of such a society, including the sponsorship of symposia on southern problems, the strengthening of the various affiliated organizations, a reduced number of meetings, the development of a conscious interest in and stimulation of research, and larger awards and grants.[21] While a few persons expressed concern that such a society, particularly when meeting in conjunction with a state academy, "would take people away from their sections and thus make the Academy meeting a kind of side show to the main attraction," most of Palmer's correspondents expressed enthusiasm for the idea of inaugurating the regional society at the Alabama academy's 1941 meeting.[22]

During this formative year, Palmer sought a wide base of support for his plan. An astute politician, he seemed to vary his expressed goals for the organization depending on his audience. For the most part his correspondence reflected the hope that industrialists could be convinced to fund scientific research. Some of his academic colleagues feared that such an orientation would do nothing for their independent university research laboratories but would only encourage businessmen to finance research that held economic promise for their companies. Palmer did not think this such a bad idea. "I believe," he wrote to Dean M. J. Funchess of the Alabama Experimental Station at Auburn University, "that the South needs *application* of research to industry and agriculture far more than academic research, at the present stage, to raise our standard of living." In time, he concluded, such progress would generate additional research funds for universities and independent laboratories.[23]

A few weeks later he wrote in a similar vein to Milton H. Fies, vice-president of DeBardeleben Coal Corporation, "that many of our richer colleges and universities in the South are apeing the North in doing academic research but are doing very little for the South." Although Fies's reply is lost, he must have chided Palmer, for the latter hastily responded: "As you state, those who are interested in applied science are deeply appreciative of the fact that pure science is the basis of all applied science, and therefore should be the first consideration of an organization like the Alabama Academy of Science. I agree with you and think the same would apply also to [the] proposed Southern Scientists Group."[24]

Prior to the scheduled meeting date of March 20, 1941, Palmer issued over three thousand invitations to scientists and businessmen throughout the South. Although the exact number in attendance was not recorded, the *Mobile Register* reported that approximately 150 persons attended the meeting. A brief description of the meeting published in the *Journal of the Alabama Academy Of Science* claimed that "delegates from practically all the scientific organizations of the South and representatives in scientific and industrial fields of eleven southern states" assembled with the Alabama academy in Mobile. Governor Dixon declined Palmer's invitation to address the group, but other notable southerners made the long trip to Mobile to speak on various topics centered on the theme "Scientific Work in the South." Appearing on the program, in addition to representatives of the Virginia, Alabama, and Georgia academies, were the noted sociologist Howard W. Odum of the University of North Carolina, who sent his message by telegram; David E. Lilienthal of the Tennessee Valley Authority, who spoke on new industries for the New South; Thomas A. Ford of the Alabama Department of Conservation, who discussed conservation in the South; Robert H. Mangum of the Alabama Power Company, who discussed defense industries in the South; James T. MacKenzie of American Cast Iron Pipe Company, who spoke on science and industry in the South; and S. R. Damon of the Alabama State Board of Health, who detailed conditions of public health in the South.[25]

Of the participants, Odum, Lilienthal, and Mangum attracted the greatest attention. Odum, director of the University of North Carolina's Institute for Research and author of *Regional Planning in the South*, claimed that "the South has an extraordinary opportunity to pioneer in the co-operation of the physical and social sciences for the development of a great region." Much of the burden, he continued, would fall on educational institutions. "Favorable attitudes and hard work are the two essentials to create in youth a new sense of the meaning of resources, standards, skills." Southern leaders must strive, he insisted, "to widen the range of occupational opportunity for youth and to train them adequately to compete with any people anywhere, anytime, to strengthen our centers of research and training, to reconstruct our agriculture and industry in the lower brackets, and to approximate more nearly equal opportunity for all people of all ages and races."[26]

David Lilienthal spelled out explicitly what he envisioned as necessary for southern development. First, he stated, "it is in the interest of the entire country that the income of the South be greatly increased above its present low level." Generating this increase, he continued, "requires a drastic change in reliance upon manufacturing or processing raw materials of the South, and less reliance upon shipping those raw materials out of the region." He insisted that the South needed an industrial development program that would focus on the resources of the region, and that "new industries for the new South must come from a partnership of the scientists, the engineer and the businessman." Anticipating the replies of skeptics and naysayers, Lilienthal offered a concrete example of the above outlined philosophy in action by pointing to a new piece of machinery that increased the yield of cottonseed oil. Designed and tested in the laboratories of TVA and the University of Tennessee, it was subsequently adapted for commercial use by firms in Georgia and Ohio, who then sold it to mills throughout the South that they might increase their productivity.[27]

Robert Mangum of the Alabama Power Company spoke to the moment when he insisted that much of the South's immediate economic woes could be solved if more defense industries could be enticed into the region. He echoed the sentiments of others that the agricultural base of the economy had to be broadened but claimed that since relatively little defense industry was located in the South, "thousands of Southern boys and girls will migrate to the highly industrialized centers, perhaps never to return—unless the present trend is changed." "Let us bear in mind," continued Mangum, "that after the war is over, many of these plants will have facilities which can be put into production for peacetime purposes. All of this will mean additional employment for skilled and unskilled men and women at good wages and will help to keep Southern boys and girls—our greatest asset—in the South." Mangum urged those scientists and businessmen in his audience to seek aggressively such industrial development, for businesses were not likely "to drift into the South of their own accord." With less in the way of manufacturing capacity than other regions of the nation, the South would have to prove its determination and potential for industrial development, and this task, concluded Mangum, should be the primary goal of the organization then being formed.[28]

In addition to the daylong symposium, the delegates attended a business session at which they formally organized the Southern Association for the Advancement of Science (SAAS) and elected Wortley Rudd of the Virginia Academy of Science as president, George Boyd of the Georgia academy as president-elect, Milton H. Fies of DeBardelaben Coal Corporation as vice-president, and George Palmer as secretary-treasurer. Palmer expressed great satisfaction with both the turnout and the results of the meeting. "We had a very fine meeting and I have certainly never before worked with a group of men more enthusiastic. All of us hope and believe, I am sure, that from the Mobile meeting, and other meetings of this organization will come some definite contributions to the South."[29]

The unanswered question remained of specifically *what* the contributions of this organization to the South would be. Judging from the sentiments expressed in the symposium, the participants wanted "progress," best defined as economic improvement through industrial diversification, leading to a higher standard of living. Coincidentally, strengthened educational facilities would entice young people to remain in the region for their training and careers, reinforcing the pattern thus established. The key, of course, was setting the plan in motion.

Rudd, Palmer, and a number of committee chairmen corresponded frequently during the subsequent year as they conceptualized the future of the organization. Hoping to attract a large and diverse membership, they outlined expansive goals which Palmer pieced together shortly before the 1942 meeting. The SAAS would strive to increase the amount of scientific research then in progress in the South, largely by pinpointing "distinctively Southern problems" and by coordinating "research efforts of educational institutions and of industry to the end that every resource may be fully utilized." To perform these functions the SAAS would encourage other scientific organizations to affiliate with it and would publish a journal "devoted to the attainment of these aims."[30] An optimistic executive committee ordered fifteen hundred membership application blanks prior to the next meeting, which was scheduled to convene on April 2, 1942, in conjunction with the Georgia Academy of Science in Atlanta.[31]

Although the SAAS was organized largely by southern scientists, many of whom hoped the organization would eventually become a southern division of the AAAS, such was not to be the case. Industrialists like Thomas Martin, Milton Fies, and Lloyd Bird of Phipps and Bird, a Richmond, Virginia, manufacturer of scientific implements, soon became the dominating force of the organization and transformed it into a regional chamber of commerce. In 1942, only a year after its founding, the SAAS changed its name to the Southern Association for Science and Industry (SASI), foretelling the trend.[32] Four years later the SASI hired a public relations expert and moved its headquarters from Palmer's office at the University of Alabama in Tuscaloosa to Richmond, the home of Lloyd Bird. Then in 1948 the SASI commissioned a fund-raiser, J. Eugene Cecil, to bring its treasury in line with its aspirations to sponsor notable conferences and publish an attractive and informative journal.[33] While the organization continued to insist that southern education needed to be improved and that the efforts of southern academic scientists deserved increased financial assistance, its primary focus revolved around enticing industry, with its research and development laboratories, into the South.

Although a number of academic scientists continued to support the SASI, others either never had any use for the organization or soon grew skeptical of its zealous boosterism. Shortly after the organizational meeting, Irving E. Gray, professor of zoology at Duke University, wrote a curt note to Palmer expressing his displeasure at finding his name on the letterhead as a member of the executive committee. Obviously appointed without his knowledge or consent, Gray asked that his name not be associated with the organization.[34]

As early as 1943 E. C. L. Miller of the Virginia academy, one of Palmer's early supporters, dubbed the efforts of the SASI as "Chamber of Commerce stuff" and complained that its primary concern had become "business and profits, first, last, and all the time. There is not one line or one thought as to who is to profit by this business or as to how profits are to be distributed. Apparently it is just assumed that they will go to the 'right persons' just as they went in the gay 1920's until one third of our people are again without adequate food, clothing or shelter or on relief."[35] By 1945 even George Boyd of the University of Georgia, long a supporter of such an organization and one of SASI's founders, bemoaned the apparent shift in orientation. "I think I recall saying," he wrote to Palmer, that "should its efforts cease to be strictly scientific and tend to become promotional in character, it will undoubtedly fail. . . . I do not think our effort is likely to proceed along strictly scientific lines if it is dominated by a group of business men." Nor was he pleased with the relocation of the main office to Richmond.

I would prefer to have the central office of the Association (if there is to be one) identified with the educational set-up of the South. I am thoroughly convinced that the scientific research program in the south cannot be developed soundly, unless the primary effort is directed toward that feature of our southern universities. I should like for our whole set-up to emphasize the place of university research, and the part which southern universities must play in the development of the region.

Boyd did not want to eliminate the industrialists from the organization. "I do not believe that the scientists can proceed effectively without the co-operation of the leaders in business." Obviously, though, he felt that the academic scientists were being pushed out of the picture, their needs overlooked.[36]

Such feeble laments went either unheard or unheeded as the SASI gathered steam. While not all southern academic scientists reacted as negatively as Miller, Gray, and Boyd, their interest in the SASI rapidly declined. By 1947 academic members were submitting letters of resignation without stating particular reasons.[37] Although occasionally the annual proceedings of the various state academies of science mentioned that one or another of their members attended the annual meeting of the SASI, formal relationships declined from supportive to tenuous at best. The SASI, as it developed a life of its own apart from the state academies that had fostered it, simply did not meet the present and pressing needs of university and college science professors. George Palmer, one of a handful of academicians able to transcend the gulf into the business world and remain unperturbed by the dominance of industrial interests in the SASI, was grieved by the lack of interest on the part of educators.[38] However, because of his conviction that academic research would ultimately profit from economic advancement in the South, whatever the method of that advancement, he could not share the concern of his colleagues.

The year 1949 marked the culmination in the shift in philosophy of the SASI and ended any hopes that even Palmer might have retained for the continued interest of academic scientists. In February of that year Palmer received a letter from H. McKinley Conway, then just launching what would become Conway Publications and editor of the fledgling *Journal of Southeastern Research*, discussing "some talk," as he put it, "of forming an Institute of Southern Engineers and Scientists." However, he told Palmer that he was of the opinion that the SASI "could probably meet the needs of the profession if some small but important changes could be made." He then proceeded to fill four pages with his single-spaced, typed suggestions for "small changes." They boiled down to a technical focus for the SASI, with increased membership dues, a permanent director, and an official (not to mention regular) publication.[39] Conway, a graduate of the Georgia Institute of Technology who had previously served in the United States Navy and as a research engineer for Ames Aeronautics Laboratory in California, found himself appointed director of the SASI before the year was out.

At first, Palmer and a few others thought that they had stumbled onto a prize. Conway provided a centralized and stable leadership that the organization had previously lacked, since elected officers were chosen annually. Furthermore, the three-hundred-dollar monthly fee that SASI agreed to pay him included distribution of the *Journal of Southeastern Research*, soon to be renamed the *Journal of Southern Research*, as the official organ of the SASI to all dues-paying members. Conway worked diligently to expand SASI membership and publicized the annual meetings well in advance in order to attract a large and diverse attendance.[40]

In retrospect, Conway was probably more interested in the long-range benefits to his publishing business that SASI contacts and the growth of southern industry could provide than he was in the future of the organization itself. In 1950 he asked for more money. SASI's budget for 1952–53 included $3,600 for Conway's salary and an additional $2,529.45 as its contribution toward publication of the journal (from which Conway also received advertising income and subscription fees from non-SASI members). By 1955 Conway had become so absorbed with managing his "publishing empire" that he suggested SASI hire another day-to-day administrator while he continued to supervise the overall operation. In that year, he founded the *Industrial South*, a journal whose title more nearly reflected his interests. Before the end of the year, this short-lived journal was merged into Conway's latest venture, the *Manufacturer's Record*, which he had purchased. Until 1958 Conway's publication continued to be distributed to SASI members. Then he decided that the *Manufacturer's Record* was "too large and impersonal" for SASI purposes and substituted the more concise and considerably less expensive *Southern Business Letter*. By 1960 Conway had bowed out of SASI completely, choosing to devote his full energies to running the now well-established Conway Publications as it cranked out texts, scientific papers, studies, research reports, and journals. [41]

In its heyday during the 1950s, the SASI published glowing reports not only of the South's industrial growth and the many features of the region that served to attract industry but also of the development of research facilities to support these industries. A number of companies, such as Navassa Guano Chemical Corporation in North Carolina, had maintained limited research facilities in the South for some time. But generally speaking, the pre–World War II "industrial South" had meant low-wage, labor-intensive manufacturing facilities such as textile plants. The economic growth that resulted from increased federal spending during and after World War II, in combination with the influx of such installations as the Oak Ridge Institute of Nuclear Studies, underlined what Palmer and others had maintained all along—that the South was an ideal region for research and development. By 1960 high technology companies had begun to consider southern locations, while private research corporations, in many instances funded by southern monies, emerged in several southern states.

Private Research Facilities

One of the earliest of these private research corporations was the Alabama Research Institute, renamed the Southern Research Institute shortly after its founding. In 1940, when George Palmer spoke to the Alabama Chamber of Commerce, Thomas Martin, president of Alabama Power Company and a member of the audience that day, pledged that his company would contribute twenty-five thousand dollars annually as seed money to found a regional research institute if others would match his offer.[42] Martin suggested that the institute be modeled after the Mellon Institute of Industrial Research in Pittsburgh. Founded in 1913 with Andrew Mellon's money and the vision of Robert Kennedy Duncan, who in 1905 journeyed to Europe to study the relationship between modern science and industry, the Mellon Institute performed specific scientific research for paying clients and also maintained a program of general research supported by foundation funds.[43]

American entry into World War II forestalled Martin's plan; eighteen months later only $5,675 had been contributed toward the research institute. Nor was the war the only retarding factor. As with the founding of the SASI, a number of people voiced objections to a private research corporation. Carleton R. Ball of the United States Department of Agriculture tried to explain to George Palmer that research institutes such as Mellon developed only after strong educational institutions had emerged in the region. Otherwise, he noted, the institutes would have to fill their professional ranks with northern scientists, thus defeating the purpose of utilizing southern talent. Dean Lee Bidgood of the University of Alabama expressed a similar sentiment when he wrote to Palmer that "I have always felt that if we were to attempt the establishment of such an institute in this or any other state at the present stage, the net result would be harmful."[44]

Palmer agreed that "reliance upon the research institutes by the industrialists most assuredly would hinder the cause of research in the institutions of higher education," but he insisted that "if the new research institutes were to work with the institutions of higher education we should have a well-balanced program." Meanwhile, Palmer and Martin remained in close contact, the former plying the latter with lists of businessmen interested in such a research institute and perhaps willing to support one financially.[45]

Undaunted by the initial lack of response to his challenge, Martin called a meeting of southeastern industrialists and businessmen to discuss the idea further. Seventy-three persons attended the meeting in December 1943, and Martin repeated his challenge offer. As a result of that meeting, thirty-five different benefactors contributed a total of $300,000.[46] The founders intended that the research institute look beyond Alabama, and in May 1944 the name officially changed to the Southern Research Institute. Begun in an old house on Twentieth Street in Birmingham, the institute launched a $2.5 million capital fund drive less than two years later. By 1947 it employed eighty persons, fifty of whom were scientists, and occupied five separate buildings. With forty-one active projects and an annual income of $320,000, the institute had become self-sustaining.[47] By the early 1980s the Southern Research Institute produced in excess of $20 million worth of research each year, generated by contracts with the United States government and private industry and supported as well by grants from foundations and private contributions. The staff of over 600 persons, including 264 research professionals and 119 technicians, publishes the results of their work in scientific journals as well as in an annual report and quarterly bulletins.[48]

Scientists in other states, while perhaps not overly enamoured of the idea embodied in the SASI, did see the research institute in Alabama as an idea worthy of emulation. In 1945 W. Catesby Jones, a chemist with the Virginia State Department of Agriculture and president of the Virginia Academy of Science, and Allen T. Gwathmey, professor of chemistry at the University of Virginia and a member of both the research and long-range planning committees of the academy, adopted as their personal project the formation of a similar institute in Virginia. As a first step they enlisted the support of Senator Lloyd Bird, who declared that the proposed Institute for Scientific Research would "be a perfect culmination of the work of the Research Committee." Then Gwathmey presented his proposal to the academy.

If Virginia is to play her part in the reorganization of the modern world, she must plan for science on a far greater scale than she has ever planned before. Existing laboratories in educational institutions and in industrial organizations, with their many diverse duties, are unable to carry the full burden of research, and new types of institutions, devoted solely to the pursuit of knowledge, must be brought into being.

In his vision, the academy would sponsor the institute and assume responsibility for raising a capital investment of $2 million.[49]

Although Gwathmey's intention of raising $2 million proved to be a bit optimistic, the Virginia Institute for Scientific Research became operational on July 1, 1947, when John C. Strickland, a biologist, was employed as a research scientist. The University of Richmond provided temporary laboratory facilities, and Senator Bird squeezed twenty thousand dollars over a two-year period from the state legislature. By 1949 the small operation had moved into the vacant Confederate Old Soldiers' Home in the R. E. Lee Camp Memorial Park, despite vigorous protests from the Daughters of the Confederacy when they discovered that Old Sorrel, Stonewall Jackson's (stuffed) horse, would have to be removed. Four years later the institute boasted sixteen employees and a budget of ninety thousand dollars. Although the Virginia Institute did not grow in proportion to its counterpart in Alabama, it remains operational in its location on the University of Richmond campus. Staffed by twenty-two persons, the institute performs research mainly in the areas of aquatic ecology and pollution effects for the United States government, the state of Virginia, and private industry.[50]

Similar organizations, some of which remain and many of which do not, emerged during this same era. Among them were the Southern Regional Research Laboratory in New Orleans, operated by the United States Department of Agriculture; the Institute of Industrial Research in Louisville, begun by the University of Louisville; and Experiment, Incorporated, in Richmond, founded by James W. Mullen, Jr., a former scientist with Monsanto Chemical Corporation, who saw a real opportunity for such a private research organization in the South. Although Mullen admitted that "basic research work is of very great importance and that without it the application of more mature technology to the problems of individual companies is impossible," the primary goal of Experiment, Incorporated, was "to be of service to the businessman, particularly in the South."[51]

By 1952 *Fortune* magazine claimed that research was rebuilding the South. "Last year," states the article, "the South sprouted one new multimillion-dollar plant each working day, for a capital accretion in excess of $3 billion." Much of this industrial surge, the magazine continued, is "largely the product of research." The article offered other supporting statistics as well, including the decline in agricultural workers from 32 percent before World War II to 21 percent in 1952 (failing, however, to note the impact of farm mechanization and the out-migration of black southerners on this figure). It pointed to 160 different corporate research laboratories and 30 independent consulting research laboratories as its final evidence. Amid this glowing report of success, though, only the Georgia Institute of Technology among the South's colleges and universities received any mention of scientific research programs worthy of note.[52]

A number of scientists expressed dissatisfaction with the emphasis on private research facilities, claiming that work performed on a contractual basis, with the client pushing for quick results, was shortsighted. Walter M. Scott, the director of the Southern Regional Research Laboratory, insisted that "too large a share of our research funds [are invested] in investigations that fit immediately into the production line, and too little in fundamental—or basic—research, which is the foundation of all science. . . . Basic research, so essential to scientific leadership, pays off in proportion to the amount of financial support it receives and the number of enthusiastic scientists who are engaged in fundamental study."[53]

Echoing these sentiments, Herbert E. Morris, a chemist for the Monsanto Chemical Company in Texas City, Texas, cautioned a group of academic scientists and industrialists attending the SASI-sponsored Southwide Research Conference in Atlanta in January 1949 that the South should not invest too heavily in private research institutes. Rather, southern industry should utilize the trained specialists in the South's institutions of higher education by contracting with university laboratories for some of their firm's needs. This procedure would save individual businesses, especially small ones, the cost of establishing their own laboratories and, more important, would serve to underwrite basic scientific research on the region's college campuses.[54]

The potential strength of a complementary relationship between scientific research on university campuses and industrial growth proved to be the foundation for the South's most famous research development undertaking, the Research Triangle Park. Nestled between the three college towns of Raleigh, Durham, and Chapel Hill, North Carolina, this park, according to historian of southern industrialization James C. Cobb, had become by the early 1970s "the site of some of the most impressive research being conducted in the entire nation."[55] The instigator of the facility was Luther Hodges, businessman and governor of North Carolina, who perceived the advantages to industry of a location within twenty-five miles of the University of North Carolina, Duke University, and North Carolina State University. Primarily because of Hodges's foresight and leadership, the Research Triangle Committee by 1957 had acquired four thousand acres of land and during the subsequent decade outdistanced other regional competitors in attracting both government research installations and such industrial firms as the rapidly expanding IBM and Burroughs-Wellcome Corporation. By 1978 the park contained twenty-two research facilities with more on the way, including the federal government's Environmental Protection Agency and the National Center for Health Statistics.[56]

Other centers, such as the University of Georgia's research park and the Virginia Science Center, developed concurrently. All of these research centers exerted a considerable impact on the local communities. By 1975, for instance, the population surrounding North Carolina's Research Triangle Park not only had expanded considerably but also could boast a greater percentage of persons holding doctoral degrees than any other area of the nation.[57] Additionally, young researchers who moved south with their employing institutions swelled the ranks of the graduate programs at nearby universities, creating a demand for new and stimulating areas of study. The result was a snowball effect, generating expanded graduate schools that in turn sent increasing numbers of men and women with masters and doctoral degrees out to work throughout the South and the nation.

Nonetheless, the South still had plenty of reason to be concerned about its academic research programs and graduate schools. Statistics from the late 1950s and 1960s reveal that the South lagged behind the rest of the nation in its support of scientific endeavor in higher education. In 1958, for example, southern institutions of higher learning spent only 20.6 percent of the national collegiate investment in research and development. Two years later, the census report revealed that the southeastern states employed only 10.9 percent of the nation's scientists and engineers, although 20.6 percent of all Americans lived in the region. Only in the agricultural sciences did the percentage of southern scientists approximate the national average.[58]

The data on graduate education were even more telling. Although graduate enrollment in the southern states, including Texas, increased 246.3 percent between 1939 and 1949 and an additional 17.4 percent during the next decade, only 16 percent of the nation's graduate students were enrolled in universities in these states by 1959. Moreover, the South produced only 14.3 percent of doctoral degrees in the physical sciences. Southern universities attracted few of the brightest students to their graduate programs, for only 15 percent of the recipients of Woodrow Wilson Fellowships, National Science Foundation Fellowships, and National Defense Education Act Fellowships chose to pursue their education in Dixie. The compiler of these figures, Cameron Fincher, director of the University of Georgia's Institute of Higher Education, concluded in 1964 that "the South still does not have a truly outstanding university," that "no university in the southeastern states appears in position to become another Harvard, Chicago, or California."[59]

Even with organizations like the SASI, private research facilities, and North Carolina's Research Triangle Park, southern scientists had been unable to fulfill their dream for university research facilities akin to those of well-endowed institutions such as mentioned by Fincher. Certainly by the 1950s considerable progress had been made, both in terms of compensated time for research and financial allotments for laboratories and travel. It was within this atmosphere of hope for continued progress that southern scientists, in the decade following World War II, assessed the need for state academies of science. As men and women increasingly relied on national organizations for career development, they once again began to wonder if the state academies had served out their purpose. Was it time now to disband? Many thought so, but others maintained that these organizations could still render a valuable service to the South and its scientists.

8 State Academies of Science in the Postwar World: Searching for an Identity

World War II changed forever the life of the southern scientist. Federal expenditures fueled the drive for a more diversified economy, and the emphasis on the importance of scientific knowledge often weighted university-wide financial decisions in favor of scientists, who had so long pled their case in vain. With an enhanced niche in the university, which had more money to invest in its human and physical resources, southern scientists found that the problems of financial stringency and geographic isolation, so common in the 1930s, no longer represented obstacles to their professional lives.

For the state academies of science, founded to combat these very difficulties, this transformation could have sounded the death knell. Instead, a few stalwart members in each of them refused to acknowledge that these societies had become anachronisms, even in the face of greatly expanded national societies and the increased opportunities for southern scientists to participate in them. Regrouping after the war, the academies created long-range planning committees to explore various avenues where they might be of service to the statewide professional communities. Overall membership did grow, primarily because (1) many more scientists had moved to the South and enjoyed the camaraderie of the annual meetings; (2) scientists and laypersons alike benefited from the special topics symposia designed to attract broad attention; and (3) academies continued to provide an avenue whereby one could be of service to the community. The journals, on the other hand, seldom commanded more than minimal respect from the scientific community.

Recovery and Growth

Recovering from the confusion of the war years proved to be the academies' first challenge. Most of them had suspended meetings during the war, and members lost touch with one another as they moved to different institutions or joined the armed forces. Membership lists became outdated, treasuries declined, and programs so optimistically approved just a few years earlier faded from view. Faced as well with stiff competition from national organizations, academy leaders soon concluded that survival depended on attracting a broader membership, offering constructive and exciting annual meetings that would encourage significant participation, somehow improving and expanding their publications, and continuing to volunteer their expertise to meet the needs and interests of the people of the states.

As early as 1940 Wortley F. Rudd, president of the Virginia academy, predicted that a new focus would be necessary if the organization was to avoid an ignominious demise. In a message to his colleagues he recounted the "almost meteoric change in both the economic and social aspects of life" that had occurred since the founding of the academy and insisted:

It will not be sufficient that we meet once a year and have a wide variety of papers, however strong they may be, on a great variety of subjects. An organization like ours may content itself with that sort of existence for the period of its youth, but will most certainly atrophy if it does not in its maturer years set itself resolutely to definitely constructive tasks that lie naturally within its sphere of influence.[1]

Wasting no time, the Virginia academy in that year appointed a long-range planning committee that represented a broad cross section of the academy membership: Ivey F. Lewis, University of Virginia professor of biology and organizer and first president of the academy; Shelton Horsley, a Richmond physician and prime mover behind the research endowment fund; Lloyd C. Bird, president of Phipps and Bird, a United States senator, and active supporter of the Southern Association for Science and Industry; Arthur Bevan, the state geologist; Virginius Dabney, editor of the *Richmond Times Dispatch*; and Rudd himself. Some newer members of the academy served on the committee as well.[2]

Rudd began by circulating a questionnaire to approximately one thousand persons, including the members of the Virginia academy, secretaries of all the state academies in the nation, and other leading scientists, in an effort to identify appropriate concerns for the state societies. Rudd categorized the responses and then listed the priorities as determined by the respondents. Not surprisingly, support for research headed the list. Other concerns that surfaced with great regularity included publicity of academy work, educational programs, state problems, science clubs and junior academies, academy meetings, national defense, and industrial problems. With the interruption of war, it would be several years before the Virginia academy could act on the results of Rudd's survey; eventually, however, it would serve as a blueprint for the academy's survival plan.[3]

The Alabama academy, admiring the initiative of the Virginia society, organized a long-range planning committee, too, just one year later in 1941. War also disrupted Alabama academy activities, but in 1944 Ernest V. Jones, professor of chemistry at Birmingham-Southern College and president of the academy, summed up the challenges he saw facing the Alabama Academy of Science. Quoting Otis Caldwell, general secretary of the AAAS, Jones told his audience that "it is imperative that there be a new age of science and society in which those who cause science to grow accept their full part of the responsibility for the proper uses of knowledge." He insisted that scientists must share their "attitude of mind" with both their students and the general public, preparing them "not only to face the facts but also to seek for all the evidence and to attempt to properly evaluate it."[4] Jones could not have anticipated more accurately the significance of such an attitude for the near future, when the United States would drop two atomic bombs on

Japan and many Americans would question the morality of the scientific research that rendered such an act of war possible.

Jones suggested several ways in which the Alabama academy could serve not only its membership but also the general public. He assumed that annual meetings and the *Journal* would continue and suggested that they should benefit a wider audience than just member scientists. He also maintained that the academy should concentrate on ways in which it could encourage research, perhaps by initiating a team research project that would benefit both scientists' professional careers and the state. Finally Jones argued that the academy should be a constant force for improving the quality of science education in the state. Citing poorly trained and underpaid teachers as the primary problem, Jones noted a recent editorial in the *Birmingham News* indicating that factory workers averaged about twice as much income as school teachers. Jones also argued that outmoded curricula should be revamped, with greatest emphasis being placed not so much on any particular subject but on instilling in students "the wish and the power to think."[5]

Obviously, Jones, Rudd, and scientists of like mind throughout the South realized that in order to survive, the state academies of science had to adapt their programs to the radically altered postwar environment. With increased participation in national professional meetings and reduced attendance at their own, all of the academies could sense the urgency for change. In no academy was this problem more apparent than in Georgia, which still maintained rigid entrance requirements and a ceiling of two hundred members. In 1947 president W. B. Redmond insisted that the academy's survival depended on a more democratic organization. With only enough papers for one session that year, the Georgia academy was obviously in trouble. In 1949 the membership, opting for permanency rather than exclusivity, abrogated the constitutional requirement of a membership ceiling and relaxed entrance standards, thus opening the way for growth.[6] —) *This doesn't really explain The need for increase in membership,*

The results were impressive, especially in light of the expanded opportunities for academic scientists to participate in national organizations. Membership jumped from 259 persons in 1949 to 349 persons only one year later. The membership roster for 1950 reveals that the new supporters of the academy were not "amateurs" who would "lessen" the high standards of the organization, but were for the most part faculty members at the state's smaller colleges who had previously been excluded from the academy because of publication requirements and the archaic membership ceiling. In addition, a number of high school science teachers joined the organization, primarily because of the academy's increased attention to secondary education through a junior academy program.[7] Obviously, educators were anxious for the opportunities provided by the state academy.

Other academies grew, too, although none replicated the sudden spurt of the Georgia organization when it removed its barriers to participation. After dropping to a low of 180 members in 1944, the Alabama academy expanded to 350 members in 1948, 459 in 1955, and 740 in 1965. It reached a high of 1,104 members in 1971. The Virginia academy, long the largest in the South, grew from 629 members in 1945 to 881 in 1955, 1,328 in 1965, and 1,767 members in 1970. Other academies expanded less rapidly, but all except the one in South Carolina at least doubled their membership between the end of World War II and 1970.

Pinpointing the exact source of the increasing membership of the state academies is difficult, because few of them after World War II recorded the institutional affiliations of their members. The Alabama academy made some effort to do so until 1961, and its roster at least partially reveals the changing nature of its membership. High school teachers and physicians associated with the Birmingham Medical Center accounted for a sizable percentage of the increased membership, but the 315 persons listing only a city address renders any sort of comprehensive analysis impossible. (See table 8.1.)

Renewed Interest in Annual Meetings

A greater understanding of this rapid growth of the state academies of science can be gained by examining the programs that attracted attention. Such growth was neither accidental nor hap-

Table 8.1
Membership in the Alabama Academy of Science, 1940 and 1961

Affiliation	1940		1961	
	No.	%	No.	%
University of Alabama	59	23.0	51	8.9
Alabama Polytechnic (Auburn)	35	13.7	50	8.7
Birmingham/Tuscaloosa[a]	37	14.5	108	18.7
Other colleges	52	20.3	49	8.5
High schools	6	2.3	76	13.2
Birmingham Medical Center	0	0	36	6.2
Other	67	26.2	207	35.9
Total	256		577	

[a]Members who listed only Birmingham or Tuscaloosa for their address; most of them no doubt had some scientific affiliation.

hazard. Concerned members realized the pull of national and specialized regional meetings and adopted a variety of tactics designed to incite the interest of potential members. Most of the academies broadened their focus to include the social as well as the physical sciences. In addition, they frequently sponsored joint meetings with other regional professional organizations, such as the state divisions of the American Chemical Society and the American Psychological Association. Joint meetings, they concluded, would relieve many persons of having to decide which meeting to attend and should thereby increase attendance for all of the participating organizations. Hoping to draw younger professional scientists as well as to maintain the interest of older members, the academies altered their sectional disciplinary divisions by breaking them into even more specialized subdivisions. While this policy often resulted in a decrease in the attendance at any one session, those who instituted it hoped that it would attract technical papers from prominent scientists by promising a small but interested and knowledgeable audience.

In 1961 E. Ruffin Jones, Jr., University of Florida professor of zoology and president of the Florida academy, addressed just this point when he urged the state's scientists to present at least a pre-view of their work at the state meeting, even if they planned a more detailed presentation elsewhere. Few if any people would hear the paper twice, he thought, and a summary review could be of great benefit to those not attending the national meeting. Twenty years later the editor of the *Bulletin* of the South Carolina academy stated his opinion on this issue even more succinctly: "Those who receive state financial support for scientific research have a responsibility to disseminate their results through an organization such as the Academy to other scientists in the state."[8]

One of the most popular and widespread methods of attracting participation in the annual meetings became the inclusion of a topical symposium, a practice long successfully utilized by the Tennessee academy. In 1951 R. T. Lagemann, president of the Georgia academy, suggested that since scientists understandably preferred to present the fruits of their major research at national meetings, the state academies "should possibly emphasize symposiums on special topics or concentrate at times on aspects of the teaching of science."[9] The Alabama academy agreed, noting that the symposium format would provide a means whereby "particular emphasis" could be placed "on the teacher, the student, and the place of science in the state of Alabama." Such a change of pace, the planners hoped, would not only relieve the tedium of technical papers but would also raise members' awareness of important state concerns.[10] As academies discovered, finding members willing to coordinate such a venture often proved difficult, and so the main feature of each meeting more often consisted of one speaker rather than an organized panel. Nonetheless such sessions proved popular enough to attract sizable audiences, including nonscientific members of the local community.

Famous names also produced good attendance, as in 1965 when Wernher Von Braun addressed the Alabama academy. A noted rocketry expert in Nazi Germany, Von Braun came to America in 1945 as the result of a United States government "manhunt" to find such scientists before Soviet Russia did likewise. Von Braun then turned his talents to furthering the American space program and spoke to a packed house. His talk was entitled "The Challenge of the Century," referring to human exploration of space.[11]

Finally, all of the academies billed their annual meetings as well as their publications as an excellent starting point for younger scholars. The old adage that one cannot get experience until after one needs it often held true in academic circles. While the presentation of papers and their subsequent publication had become a requirement of professional life, doctoral candidates and those with newly minted degrees frequently found their articles rejected by major journals in favor of the work of more noted scholars. The state academies, though, offered these young professionals the opportunity to "break the ice," representing a viable alternative to the often crowded programs and selective journals of larger societies.

One other aspect of these postwar annual meetings should not go unnoticed, and that was a return to an older stated purpose of scientific societies, namely the diffusion of knowledge to the general public. The symposia, often held in the evening, were designed for this purpose. The academy also considered the presidential addresses as a means for scientists to reach out to the public and dispel some of the mystery and fear that surrounded scientific research since the dawn of the nuclear age. Whereas prior to World War II most of the presidential addresses had centered on the speaker's own research for an audience of his peers, those in the postwar years concerned more general topics and sought a broader audience. The use of special guest lecturers, such as Wernher Von Braun in 1965, was also intended to appeal to the general public as well as to the membership.

As always, the academies seldom kept attendance figures at any of the sessions. Infrequent records do indicate high participation, though, especially in academies that had been strong prior to World War II. At its annual convention held in Durham in April 1975, the North Carolina academy attracted an estimated 1,000 persons, including students and members of affiliated societies meeting concurrently. The Virginia academy had always drawn good attendance, and this success continued. By 1949 registered attendance had mounted to 783 regular members and 248 members of the junior academy. Participation dropped slightly during the 1950s, averaging between 500 and 700 persons, about one-third of whom were not members of either the senior or junior academy. But the mid 1960s, though, total attendance climbed over 1,000 persons, where it has remained. The number of papers presented at VAS meetings is similarly high; at its fiftieth anniversary celebration in 1972, 467 papers plus 4 symposia filled the program for the three-day meeting.[12]

Not all of the academies enjoyed such success as that evidenced in North Carolina and Virginia, and reflected also in scattered attendance figures for the Tennessee and Alabama organizations. The South Carolina academy, never a strong one, reported only 204 regular members in 1975; one year later, a scant 132 persons attended the annual meeting. The Louisiana academy has fared only slightly better. Unfortunately, published academy records are so haphazard that it is impossible to determine even the number of active members on an annual basis, let alone attendance at the yearly meetings. Available lists do show a slight growth trend, from 135 members in 1935 to 273 by 1949, 361 in 1967, 464 in 1971, and 425 in 1977.[13]

Minority Participation

All of the academies overlooked (or ignored) a significant and growing pool of potential members by failing to include interested and qualified blacks in their ranks. During the 1930s a few members of the Alabama academy had discussed inviting George Washington Carver of Tuskegee Institute to their meetings, but they never did so. In 1944 academy president Ernest V. Jones suggested to the membership that they either encourage black scientists to organize their own academy, form a separate section of the Alabama academy for blacks, or admit them to the organization as regular members. Earlier that year Jones had corresponded with E. C. L. Miller, secretary of the Virginia academy, concerning this very issue. Miller informed Jones that his membership records did not indicate a person's race, and so he could not state for certain if any academy members were black. He mentioned that from time to time "colored persons" had presented papers before the academy, although he failed to cite specific instances. Miller did concede that social functions had created a bit of a problem, "but in recent years we have been getting away from such occasions."[14]

Miller continued to ponder this issue, and two weeks later the Virginia scientist wrote to Jones with additional comments. Miller had discussed the matter with Garnett Ryland, professor of chemistry at the University of Richmond and a leader in Virginia's interracial commission, whose advice was just to "go along and treat the colored people just as you would treat anyone else and not do much talking." Miller concluded to Jones that "whatever you do or do not do it will be a long and slow process. . . . I know of no white person here in Richmond who is disturbed in the slightest because the railroads charge colored people first class fares and give them second class service. . . . Everyone just goes blythly along and completely ignores it. Against such a smug moral vacuum I fear yours would be a voice crying in the wilderness."[15]

Miller discussed the situation with other members of the academy too. Dr. Sidney Negus, professor of chemistry at the Medical College of Virginia, seemed surprised that "Alabama is so far behind the times that the Academy of Science still does not accept Negro members." He was of the opinion that the Alabama academy should open their membership to blacks "*before* they are eventually pressured into it. Science should take the lead in helping with the racial problem." Miller agreed, commenting that it would indeed be humiliating when the Supreme Court decided to tell the South how to behave.[16]

Although members of the Virginia academy, in correspondence with one another and with scientists in other states, sounded as if the racial issue had created few problems, their perspective was more typically southern and paternalistic than they would have wanted to believe. The suggestion to treat blacks like everyone else and "not do much talking" smacks of an unwillingness to rock the boat in the face of injustice. It also suggests a lack of understanding of the difficulties faced by black members who wanted to participate in annual meetings. While sessions most often took place on college and university campuses, out-of-town members still needed a place to stay and dining facilities. In addition to the problem of obtaining adequate accommodations, black participants sometimes found themselves excluded from academy social functions held in public hotels and restaurants. In 1949, several years after the above discussions, Lubow Margolena Hansen wrote to Foley Smith, a chemist with the Virginia Alcoholic Beverage Control Board and president of the Virginia academy that year, asking if "an attempt has been made to arrange the Yearly Meetings of the Academy at a place where all members could meet and visit without any embarrassment." A handwritten note attached to the letter identified Hansen only as a black woman.[17]

Despite the rhetoric, then, southen state academies of science did little to promote racial justice. In 1944 E. V. Jones, subsequent to his correspondence with Miller, told his Alabama colleagues that "educated negroes in the South are presenting many new problems." An interesting choice of words, and probably not one that Jones, even in retrospect, would consider inappropriate. No state academy of science accepted the challenge of leadership on this issue. In 1955 the executive committee of the Alabama academy voted to admit qualified blacks to full membership, but it stipulated that "the Academy would have to tell him [the black applicant] in all sincerity and sympathy that he would not be able to participate fully in all the activities of the Academy, on account of local ordinances beyond the control of the Academy." Other southern academies of science quietly opened their organizations to black scientists at about this same time and in much the same manner.[18]

Jones's attitude toward women scientists was similarly paternalistic, reflecting a situation that has been well documented by historian Margaret Rossiter. She claims that between 1880 and 1910, the entry of women scientists into traditionally masculine occupations threatened male scientists and "resulted in the women's almost total ouster from major or even visible positions in science." Although admitted to the profession, they were relegated to subordinate positions in the name of "higher standards." Later, qualified female scientists who tried to scale these barriers encountered the argument that "no precedent" existed for their promotion, and thus they faced the stereotypes that precluded their full participation in professional societies and their complete access to the job market.[19]

All of the southern academies admitted qualified women, but they represented a small minority of the membership and seldom held leadership positions. By 1928, for instance, the North Carolina academy had grown to 247 members, only 37 of whom were women. In 1933 Helen Barton became the first female academy officer upon her election as vice-president. Not until 1941 was another woman elected to office, when Eva Campbell, a professor of biology at the North Carolina College for Women (now the University of North Carolina at Greensboro) and a member of the academy since 1921, was elected vice-president. The following year Mary E. Yarborough, professor of chemistry at Meredith College, was chosen as the academy's vice-president. In that year members of the nominating committee noted that while women composed about 20 percent of the membership, only the three just mentioned had ever held office and that the committee should seek women for academy offices.[20] In fact, however, they did no such thing. During the subsequent two decades, only four women held elective office in the academy.

Prior to 1960 the Alabama academy elected only two women to office. Septima C. Smith, professor of zoology at the University of Alabama, served as academy secretary from 1935 to 1940, and Winnie McGlamery filled the same office from 1940 until 1946. In 1944 E. V. Jones, in his address to the Alabama academy concerning challenges that the organization faced, noted that "the effects of the manpower shortage are clearly evident in our meeting." He concluded that the absence of male scientists provided an opportunity "to the women workers in the various fields of science in Alabama for a larger participation in the activities of the Academy." Reminding his audience of a recent film on Madame Curie, he stated that he recognized there "a quality of mind that is coming to be recognized as playing a very important part in expanding the frontiers of knowledge." "Intuition," he continued, "has long been regarded as a peculiar quality of the feminine mind." He then noted that Irving Langmuir, a recent Nobel laureate, had himself insisted that intuition had played a significant role in his own research discoveries. For Jones, this statement was sufficient evidence that the "peculiar" female quality of intuition could perhaps contribute to scientific endeavor.[21] Women scientists of the postwar era faced an uphill battle.

Expansion of Journals

Excepting blacks and women, southern state academies of science did succeed in increasing their membership among people who had not previously supported the organizations. The nucleus of their membership, however, remained college professors, and in order to retain the interest and loyalty of this population the academies had to offer more than just an annual meeting. Well aware of the increasing pressure on academic scientists to publish the results of research efforts, stalwart members of these academies insisted that every effort should be made to improve their journals. Timely publication and a true journal format—as opposed to abbreviated bulletins—would, they reasoned, attract both a greater number and a higher quality of articles. While even such expanded journals would never rival those of national professional societies, they could fill a void by publishing very specialized articles and the efforts of younger scholars. Naturally, journal editors encouraged all members to contribute and hoped occasionally to publish an article that would receive national attention.

The *Journal of the Elisha Mitchell Scientific Society* and the *Journal of the Tennessee Academy of Science*, the only two publications of southern state academies of science worthy of calling themselves a journal before World War II, continued much as before. Both appeared quarterly, if sometimes a bit late, and contained proceedings of the annual meetings of the academies, a varying number of articles, abstracts of papers not published in full, and reports of academy committees. Neither academy ever received regular state funds to support its journal or the use of the state printing office, although the *JEMSS* was published through the auspices of the University of North Carolina Press. Compared to national publications, they were not particularly strong.

In 1973 the North Carolina journal was in such dire financial straits that it was saved only when the university agreed to meet all deficits through that year. It then instituted a thirty-dollar page charge payable by the author. The unpopularity of this policy was evidenced by the report of the editor, who in 1978 complained of the dearth of good manuscripts submitted for publication.[22] Furthermore, by 1983 the journal had fallen two years behind schedule. This haphazard publication schedule and the high page charges have not enhanced the journal's reputation, which remains well below that of national organizations. In the late 1940s, though, the *JEMSS* was held in high esteem by the other southern state academies of science, whose publications amounted to little more than brochures detailing the annual meeting.

The Virginia academy, the largest and in many ways the strongest of these southern societies, was determined to publish a journal that would compare favorably with those of similar organizations. The first three issues of the *Virginia Journal of Science* appeared in 1940, 1941, and 1942; then financial difficulties and the uncertainty of the war years brought its publication to a temporary halt. The *Journal* did not resume its schedule until 1950, for, as E. C. L. Miller remarked, "Like all organizations the Academy is having trouble adjusting to our American dollarettes."[23]

Boyd Harshbarger, a statistics professor at Virginia Polytechnic Institute in Blacksburg, accepted editorship of the *Journal* in 1953 and entertained high hopes for its success. He expanded the staff to include a technical editor, an assistant technical editor, and an advertising manager. Section editors, usually academy members serving on a volunteer basis, were to provide assistance in their specialized fields and thus relieve some of the burden on the staff. Harshbarger envisioned the *Journal* as a quarterly that would contain the proceedings of the annual meeting, a membership roster and other pertinent information, and five or six scholarly articles in each issue.[24]

Despite his early enthusiasm, Harshbarger suddenly resigned in 1955, citing increased duties at VPI. Frustration was a more probable cause. The new editor, Horton H. Hobbs, Jr., expressed displeasure with the small backlog of articles awaiting publication and echoed the sentiments of his counterparts elsewhere by pleading for more and better papers. Meanwhile, mushrooming publication expenses consumed the academy's income and part of its reserve fund. A disillusioned Hobbs resigned four years later. In 1960 the academy agreed to increase its subsidy to the *Journal* and to allow additional advertising. Evidently these measures did little to improve the situation because Marion B. Ross, Hobbs's replacement as editor, resigned in 1961.[25]

The *Virginia Journal of Science* never achieved the status of which Harshbarger and others had dreamed. The major impediment, of course, was financial. Although the Virginia academy maintained a large trust fund, the income could be used only to support research grants. A less obvious problem may have been Harshbarger himself. Following his resignation as editor, he remained on the academy committee that oversaw the operation of the *Journal*. An outspoken man, he no doubt expressed his dissatisfaction with the quality of the quarterly publication. In early 1961 the committee outlined a production schedule for Ross to follow, suggesting that if he could not do so, he should resign. As a result, the *Journal* had to appoint its fourth editor in seven years.[26]

The size of the *VJS* declined markedly during this time. From a high of 490 pages in 1958, it fell to 262 pages in 1960. Abstracts of papers presented at the annual meeting consumed almost half of the *Journal*, leaving relatively little space for complete articles. The *Journal* continued to appear quarterly, although some issues contained as few as 40 pages. In 1973 Perry C. Holt, chairman of the publications committee, made the all-too-familiar plea to members of the Virginia academy to support the *Journal*. Holt's editorial reveals not only the difficulties faced by the *VJS* but those common to all state academy publications by this time.

Basically, claimed Holt, good articles were too few and far between. "Senior scientists within the Academy," he wrote, "must be encouraged—urged—to publish more frequently some of their best work in the *Journal*." Historically, these scientists have opted to publish in national professional journals "for the sake of greater recognition of their work and the enhancement of their reputations nationally and internationally." Also, he stated, "the administrative officers of at least some of the institutions of higher education in the state have actively discouraged their staff from publishing in the *Journal*." In an attempt to rebut this argument,

Holt insisted that his own articles in the *Virginia Journal of Science* "have been as widely cited as those published in national or international journals." He pointed to the academy's large journal exchange program, which carried the *VJS* throughout the nation and the world, as evidence enough that it should be considered a viable publishing outlet. In conclusion, he maintained that "the advantages of speedier publication and lower costs help to outweigh the objections that have been raised."[27]

Holt might as well have saved his energy. In 1978 the editor complained of inadequate clerical help, recalcitrant authors, and too few good articles.[28] Although the *VJS* of the 1980s comprises approximately three hundred pages in four annual issues, one issue, usually the largest one, is devoted completely to abstracts of papers presented at the annual meeting. In an effort to keep expenses manageable, authors must provide camera-ready copy, a publication procedure that detracts from the physical appearance of the *Journal*. Authors are also required to pay a charge of twenty-five dollars per page if their printed work exceeds fifteen pages.

The journals of the other southern state academies have had similar experiences. All of them expanded in the years immediately following World War II, when membership grew rapidly and the future semed bright for their continued growth. But the pull of national organizations was too strong for the state academies to compete against. The journals suffered as a result, both from an ever-pressing lack of funds and from a dearth of quality manuscripts. Journal editors, although they perennially conclude their annual reports to the membership with a plea for more papers representing major research, have for the most part accepted the limited role that their publications play in the professional lives of modern scientists.

Not to be undermined completely, they have followed the lead of those who organize the annual meetings and frequently publish articles relating to matters of statewide interest, particularly those that grow out of the academy symposia. The summer 1984 issue of the *Journal of the Elisha Mitchell Scientific Society*, for instance, commemorated the centennial of the founding of the Mitchell Society by devoting its pages to articles relating to Mitchell himself, the founding of the society, and the historical significance of the journal. All of these papers grew from a joint meeting of the Mitchell Society and the North Carolina academy in April 1983, at which time the Mitchell Society formally "passed the torch" to the academy and disbanded. The journal of the Tennessee academy has long followed such a practice, publishing a myriad of articles generated by research at the Reelfoot Lake Biological Station as well as those centering on such issues as the Tennessee Valley Authority and the Oak Ridge Institute for Nuclear Studies. Much of the Virginia academy's research concerning the potential of the James River Basin appeared in the pages of the *Virginia Journal of Science.*

The modern journals of the state academies of science remain relatively small and usually require some form of author subsidy, such as page charges. Some journals stave off increased publication costs by combining two of their quarterly issues into one larger issue and by accepting only camera-ready copy from authors. A few of the journals appear only annually. While these publications certainly are not what the academies of the 1940s had hoped for, they should not be dismissed as insignificant. The abstracts of papers from the annual programs serve notice of research in progress; the chance to publish a first article is crucial to young scholars; symposia papers offer readers an opportunity to be informed of local and regional issues. In short, none of the academies has opted to save money by discontinuing the publications. Having abandoned the dream of rivaling national publications, they have found a niche that they can comfortably fill.

Other Professional Concerns: Regional Issues and Research

During the 1950s the academies attracted both attention and members with the broadened focus of the annual meetings and the somewhat expanded journals. Social scientists and historians of science joined their ranks, and all manner of scholars found the accessibility of their annual programs and the pages of their journals an occasional alternative to crowded national conventions and selective scholarly publications. This diversity, however, triggered by the increase in the number of sectional divisions, acted as a centrifugal force on these state organizations. Indeed, with sections that annually elected chairpersons, vice-chairpersons, and editors, some academies appeared to be little more than a loose confederation of specialized scientific societies that agreed to meet at the same time. Poorly attended business meetings only enhanced this impression.

The force that most often held the various sections together, other than a common treasury, was the interest generated by standing committees, such as those on conservation and education. These matters cut across sectional interests and touched the private as well as the public lives of all the members. Wise use of state resources had always been a major issue faced by the state academies, and matters of conservation have continued to attract their attention. The conservation committee of the North Carolina academy has remained especially active, serving primarily a watchdog function. In 1952, for instance, it warned that members should be alert to any attempt to undo the legislation banning the exportation of Venus's-flytrap plants from the state, hinting that Wilmington dealers might very well try for a repeal. The academy's efforts in 1953 and 1954 to save Crabtree Creek State Park from becoming a United States Air Force base have previously been noted. Nor was the academy willing to allow any other sorts of tampering with the wild public domain. While they agreed that national and state forests should serve as multipurpose preserves, meant not only to conserve flora and fauna but to provide for human recreation as well, they regularly opposed efforts to dam up streams that would create recreational lakes but destroy in the process wild habitats; they joined a chorus of outcries against legislation that would place control of grazing rights on federal lands under local control, claiming that local officials were too often themselves controlled by the cattle interests; and they lobbied against a proposal from Central Telephone Company that the company be allowed to erect a microwave tower in Mount Jefferson State Park.[29]

In some cases, academy conservation efforts met with success, as with the relocation of Seymour Johnson Air Force Base and the defeat of the Stockmen's Bill. They also won the receptive ears of state lawmakers with their pleas that part of the barrier islands, known as the Outer Banks, be placed under state control and preserved in their natural state. On the other hand, their opposition to a new four-lane interstate highway through the Smoky Mountains went unheeded.[30] Evidently few members of the academy had ever made the trek from Asheville, North Carolina, to Knoxville, Tennessee. The hundred-mile journey could take as long as four hours if travelers found themselves following a tractor-trailer as it wound its way along the two-lane road. Nor does modern road construction necessarily mean destruction of everything in its path. Interstate 40, as it traverses the mountains and for a stretch parallels the French Broad River, provides breathtaking scenery to travelers, with very little intrusion into the landscape.

The Virginia academy likewise continued its attention to conservation. In addition to watchdog functions, the committee occasionally undertook studies of important regions of the state. Having previously labored to save the Great Dismal Swamp from real estate developers, the Virginia organization sanctioned a detailed study of the wildlife of the region. Beginning with the 1969 issue of the *Virginia Journal of Science*, components of this study reached the public attention. The Virginia academy also expressed concern over the manner of use and control of federal lands, an issue that attracted increasing public attention during the 1970s and 1980s.

Most of the Alabama academy's concern for state resources centered on the individual, as it constantly spoke out in favor of improved science education from the primary grades up through high school. As with other similar organizations, much of the academies' clout was spent when they had aired their views in annual symposia and the journals and perhaps urged concerned individuals to write their representatives in Congress. The academies alone could not accomplish all that they envisioned. Rather, success depended on additional support and, most significantly, on the lawmakers' perceptions of the desires of their constituencies.

Another professional need that the state academies of science had struggled to meet from their inception was that of research funds for their memberships. Before World War II, their attempts in this regard were limited at best, and although they continued many of the research grant programs, the stipends thus awarded did not increase appreciably. In addition to the small sums received annually from the AAAS and some private donations, the only other source of grant money for the academies came from the National Science Foundation, and the availability of such federal money fluctuated considerably from one year to the next.

Compared with grants from other sources, the academy grants to scientists were minimal. For example, in 1974 the North Carolina academy honored thirty-two of fifty grant requests, for a total of $2,054. Recent reports of the Alabama academy's research committee do not include sum totals but do note that usually out of eight to ten grant requests, six or seven receive some funding. Likewise, requests for grants remain low in the Virginia academy. In 1980 the research committee awarded six proposals a sum total of $2,585.[31] Obviously, in this age of five- and even six-figure grant proposals, southern scientists do not rely on the state academies of science for the financial support that they need.

Thus the state academies of science that had hoped, during the postwar years, to rekindle interest and participation in their organizations often found their aspirations thwarted. While southern scientists' entry into the national scientific community exerted a positive effect on regional development, the impact on the state academies was quite different. Having been organized at a time when southern scientists desperately needed the contact that such regional societies could provide, the academies suffered what can best be termed an "identity crisis" as their present and potential members turned to the more specialized meetings and prestigious journals of national societies to fulfill their professional needs. In an attempt to reverse this trend, the academies broadened the scope of the annual meetings, struggled to enhance the quality of their journals, continued the admittedly small research grant programs, and addressed such specific state interests as conservation of natural resources.

These efforts met with limited success at best. Most effective in terms of attracting attention to and participation in academy affairs were the rejuvenated annual meetings, which combined very specialized sessions with symposia addressing issues that cut across disciplinary divisons. Young scholars, including graduate students, often feel more comfortable in presenting their first papers at these meetings rather than in the more intimidating atmosphere of national conventions. More established scholars sometimes utilize this forum as well, and in addition enjoy the camaraderie with their colleagues. Laypersons frequently attend the symposia, which often focus on issues of statewide concern, and thus scientists find themselves able to reach the general public on matters where their expertise is a valuable asset.

The hopes for expanded journals and significant research grants were never to be realized, though. Funding remained a severe problem, as publication costs and grant proposals skyrocketed with general inflation. Membership dues could not begin to cover the increased expenses. Industrial memberships, averaging about one hundred dollars per company annually, never materialized to the desired extent; nor did state support provide the continued regular funding on which the academies could depend as they planned for the future. Consequently, journals and grant programs remained small, unable for the most part to attract either the necessary funding or the high-powered articles and proposals that would generate significant expansion. Eventually the academies reduced their aspirations to match their financial capabilities. The journals serve as "bulletin boards" for academy affairs and a publishing outlet for younger scholars, occasionally garnering an article of national importance. Likewise the research grants best serve the needs of those whose financial requirements remain modest.

Given the limited success of the state academies in meeting professional needs of southern scientists, their membership spurt of the postwar era seems puzzling. Of course, much of the growth can be accounted for by the sheer increase in the number of scientists in the South as colleges expanded their faculties and industry moved into the region. Younger scholars especially viewed the academies as a perfect starting point to present their research and gain a modest publication credit. Another significant component of the academies' growth, at least in some states, resulted from the active recruitment of high school science teachers. Then too, academy membership was not especially demanding; dues remained low, and organizational tasks could be avoided altogether. Consequently, it cost scientists very little, financially or timewise, to join these organizations. And there were some benefits of membership. Scientists frequently cited the camaraderie provided by the state meetings as one reason for membership; also, the modest journals often contained notice of scholarship underway in the form of abstracts, as well as occasional interesting and significant articles. For those men and women willing to share their expertise and speak out on regional issues, the academies provided a forum, such as the ongoing conservation committees. In the postwar era, an issue of great concern to scientists everywhere, the South included, was science education at the secondary level. The effort of southern state academies of science to maximize the region's human potential in this area was one of their greatest contributions and deserves detailed attention.

9 Educating the Next Generation: The Academies' Role

By the 1950s most of the southern state academies of science had created a niche for themselves in the professional lives of many academic and some industrial scientists, as evidenced by the increased membership. While the journals proved to be a disappointment, the academies did offer camaraderie through the annual meetings and an opportunity for those who so chose to serve on committees that addressed such particular regional issues as the environment and education. In fact, by 1950 most of the academies had identified science education as their chief service priority. As we have seen, scientists' concern for the quality of secondary science instruction was not new. During the two decades following World War II, though, the perceived crisis of a national shortage of scientific personnel generated a new urgency. Reacting to the widespread fear that the United States might fall behind the Soviet Union in technological expertise, southern state academies of science, along with their counterparts throughout the nation, established lines of communication with state departments of education, organized junior academies of science, initiated scholarship programs, cooperated with the Westinghouse Foundation's national science talent search, sponsored science fairs, and participated in visiting scientist lecture programs. While the height of such activity occurred during the 1950s and 1960s, many acade-

184

mies have maintained various science education programs since their inception.

Scientists were not the only Americans in the postwar world to demand an improved school curriculum. The technological explosion that accompanied the Second World War and the resultant need for highly skilled personnel forced legislators to take heed of what educators had been saying for decades. The Soviet Union's successful launching of Sputnik in 1957 added fuel to the fire as Americans became determined not only to catch up but to excel technologically. Educational programs that had previously suffered for lack of financial backing suddenly found a wealth of contributors. State industries volunteered funds to academies of science to sponsor junior academies and science fairs; the National Science Foundation supported a variety of programs administered by the academies; universities offered scholarships to winners in science fairs and talent search competitions; and the American Association for the Advancement of Science altered its research grant policy to insist that the money be awarded only to students. With this new national priority, state academies exuberantly trotted out new plans and old, and for the first time found financial supporters waiting in line.

Most of the southern state academies of science had at least addressed the issue of science education by 1940, so that following the war they quickly picked up where they had left off. Since very few of them enjoyed the close relationship with the state board of education as that maintained in North Carolina and detailed previously, they turned to direct contact with teachers and students as a means of stimulating interest in science and improving the quality of instruction. In many cases they discovered a ready-made base in local high school science clubs.

Junior Academies of Science

In 1933 the Alabama junior academy had begun in just this fashion, organizing a state meeting of high school students through the science club network. Local clubs affiliated with the junior academy and thereby gained the right to choose official delegates to attend junior academy council meetings. Each club could send as many representatives to the state meeting as they chose, and the organizers encouraged all young people to present papers or prepare a visual exhibit. The senior academy maintained close supervision over the junior academy through two counselors appointed annually and, beginning in 1950, a permanent counselor.[1]

The Alabama academy found a ready ally to help publicize the junior academy in Science Service Incorporated of Washington, D.C. Organized in 1920 in response to public demand for more complete coverage of scientific news, Science Service translated complex issues into laypersons' terms. During subsequent decades the organization widened its scope to provide a similar service to secondary schools, both teachers and students, through Science Clubs of America. This branch of Science Service, already in touch with many of Alabama's high schools, encouraged the local science clubs to become a part of the state junior academy. For a small fee the junior academy counselors received an annual handbook and other literature pertinent to science club advisers, as well as information concerning the National Science Talent Search and national science fairs. By utilizing this service, the Alabama academy enjoyed considerable success in spreading the junior movement beyond the urban areas of northern Alabama.[2]

In his academy presidential address of 1946, J. M. Robinson, a professor of zoology at Alabama Polytechnic Institute in Auburn, encouraged his Alabama colleagues to lend their assistance to this effort, reminding them "that the trained minds of the youths of Alabama will help to develop the sciences in this great state."[3] Several members responded to the call. The untiring efforts of James L. Kassner, a University of Alabama chemistry professor, as well as high school teachers Clustie Mctyeire and Kathryn M. Boehmer, sustained the junior academy through the early years of uncertainty. In 1945, aiming toward expansion, Boehmer divided the state into thirteen districts, appointed a director from the senior academy for each region, and charged those persons with the responsibility of maintaining contact with all high school science clubs in the area. She also mailed a bulletin detailing junior academy activities to three hundred white high schools in the state.

The following year the junior academy issued its first *Alabama Science News* to all members. By 1959 the Alabama junior academy had grown so large that it subdivided into four regional academies and began a fifth for black students, each of which held its own annual meeting. Although any student could attend the annual statewide meeting, the demand to appear on the program became such that participants competed for the honor at the regional level.[4]

The junior academy movement, already well established by the 1940s in the states of Illinois, Indiana, Kansas, Iowa, Kentucky, Pennsylvania, Alabama, Oklahoma, Texas, and Minnesota, spread so rapidly that Hubert J. Davis, chairman of the Virginia Junior Academy of Science Committee, stated enthusiastically, "There is nothing parallel to this movement in the history of science."[5] Academies that had debated the idea of a junior academy throughout the 1930s now quickly jumped on the bandwagon. By 1947 every southern academy of science except North Carolina sponsored a junior academy, although the one in South Carolina soon fell dormant.

Amazingly, the southern academy that led the way in calling for improved science education did not found a junior academy until 1974, although it had sponsored a collegiate academy since the mid-1950s.[6] In 1931 the North Carolina academy agreed to support high school science clubs but did not recommend their affiliation into a junior academy because of the "uncertain life" of such clubs and the increased responsibilities that would fall upon the secretary of the organization.[7] In 1935 C. F. Korstian, professor of forestry at Duke University, broached the matter again after learning of the Alabama junior academy, but he concluded that stimulating high school science clubs "seems preferable to sponsoring the so-called junior academies."[8] Perhaps the North Carolina scientists were satisfied that their efforts through the state board of education, state teachers' organizations, and the various essay contests met the educational needs addressed elsewhere by the junior academies. Beginning in 1959 the North Carolina academy did inaugurate sponsorship of an annual Junior Humanities and Science Symposium in conjunction with Duke University. The academy also agreed to cooperate with other state organizations in sponsoring state science fairs and an annual science talent search and thus reached high school students in a variety of ways.

Most of the southern junior academies that sprang up in the late 1940s developed in a fashion similar to the one in Alabama. Student representatives met annually in conjunction with the senior academy, elected officers for the year, heard papers by their fellow students, and prepared exhibits for competitions. The growth of any one junior academy occurred in direct relation to the dedication of the teacher-sponsors at the local level and the coordinators in the senior academy. In Virginia, for instance, the junior academy did not always run smoothly, for few teachers volunteered to assume the extra responsibility, school administrators did not seem to grasp the importance of the movement, and senior academy members for the most part did little more than talk about the need for such an organization, evidently hoping that "someone else" would translate ideas into action.[9]

National Science Talent Search

In addition to junior academies, the inauguration of the Westinghouse Science Scholarships, funded by the Westinghouse Electric Corporation, spurred the interest of senior academies in high school science students. In the fall of 1941 representatives from Westinghouse's Educational Foundation approached Science Clubs of America (SCA) with a proposal to seek out and provide for the education of high school seniors who exhibited superior scientific talent. In the program as established through the offices of SCA, applicants completed a three-hour science aptitude examination and submitted letters of reference from high school teachers, their transcript, and a one-thousand-word report outlining a project that they were prepared to present. Word of this competition passed from SCA to senior academies, junior academies, science clubs, and local school boards.[10]

Although some state academies of science took scant notice of the National Science Talent Search during its early years, others by the mid-1940s realized to their chagrin that few southern students entered the competition and that none of them had won one of the forty scholarships awarded annually. A 1945 report of the North Carolina academy's Committee on Science Education pointedly remarked that "a recent analysis of the National Science Talent Search places North Carolina practically at the bottom."[11] This information came from a 1944 article in *Science Education*, "Is Your State Discovering Its Science Talent?" and was intended to challenge those states with a below-average rate of participation in the talent search program.

Not surprisingly, every southern state ranked below average in the number of student participants per 1,000 population. For instance, both Wisconsin and North Carolina had approximately 35,000 high school seniors in 1940, but 433 Wisconsin students entered the contest, as contrasted with 70 students from North Carolina. In the first four years of the program, Wisconsin sponsored 45 students who won honorable mentions and 9 who placed among the final forty winners. North Carolina, on the other hand, had sponsored only 2 honorable mentions and no finalists. While Florida students exhibited the highest participation rate among the southern states, with 2.8 entries per 1,000 high school seniors, the other southern states were grouped near the bottom with between 1.8 and 0.7 entries. By contrast, sixteen states in New England and the Midwest had over 3.0 entries per thousand, and eight more ranked between 2.5 and 3.0.[12]

Several southern state academies of science determined to improve their region's showing in the talent search competition. Jacob W. Shapiro, a high school teacher in Columbia, Tennessee, encouraged that state academy to sponsor a statewide talent search in conjunction with the national one. In a paper presented to the Tennessee academy in 1945, Shapiro noted that only six Tennessee students had received the honorable mention award in the national talent search contest during the preceding four years. In 1945 Tennessee had its first finalist, Alice Beck Dale, a student at Central High School in Columbia. Pointing out that 120 science clubs operated in high schools throughout the state and could serve as a nucleus for statewide organization, Shapiro suggested that all students who entered the national competition should be eligible for the state one, which would be judged by an academy committee. With a bit of legwork and persuasion, he concluded, academy members could persuade colleges and universities throughout the state to offer scholarships for state competition winners, thus making the program more attractive to students. The Tennessee academy accepted this challenge and was gratified by the results. Five years later, the academy judges had recommended twenty-two finalists for scholarships at institutions throughout the state, and one student from Oak Ridge had won second place in the national competition.[13]

The Virginia academy adopted exactly the same plan in 1946. Science Clubs of America agreed to send to the academy the entry information from all Virginia students, just as it did for the Tennessee academy. Using these data, judges from the Virginia academy selected state winners. Although these students received no monetary reward, the Virginia academy invited them to read their papers at its annual meeting, encouraged them to apply for scholarships at the college or university of their choice, and agreed to write letters of recommendation on their behalf.[14]

"Since one of the main objectives of the Virginia Science Talent Search was trying to better teaching in the high schools of Virginia," stated the director of the Virginia program after its first year, "all secondary school principals and members of the State Department of Education were sent full information of the results of the search." This action produced a tremendous increase in the number of student participants. By 1948 a total of 146 Virginia students had entered the national contest, and 4 of them won honorable mention. That year, 15 winners in the state competition divided one hundred dollars in prize money. The committee pleaded with the academy for more help, more prize money, and some suggestions for obtaining scholarships in Virginia for these students. Although the academy treasury did not respond, twenty-eight colleges and universities throughout the state began in 1952 to offer over one hundred scholarships to participants in the program, and industries contributed about thirteen hundred dollars to defray the cost of the competition. Also in that year, Virginia produced its first national award winner, Ruth Flinn Harrell of Norfolk.[15]

In 1947 the Alabama academy also initiated a state science talent search. The members were committed not only to increasing the participation of Alabama students in the national talent search but also to ensuring the continued education of talented students in the state. Science Service agreed to cooperate with the Alabama academy in much the same manner that it did with the Virginia and Tennessee organizations, and following the national competition of 1947, all records submitted from Alabama students were turned over to academy officials for the state competition.[16]

Not wishing to leave the matter of scholarship awards to chance, James L. Kassner, a University of Alabama chemistry professor who had been a vital force in organizing the junior academy, asked several major colleges in Alabama to commit themselves to granting one four-year tuition scholarship for each of the following four years to a winner of the state talent search. The University of Alabama, Alabama Polytechnic Institute, Birmingham-Southern College, Howard College, and Tuskegee Institute all agreed to do so. Obviously, the Alabama academy intended to include black students in the competition, for otherwise they would not have approached Tuskegee Institute. In addition, Kassner broached the idea to a number of leading industrialists of the state, including Thomas W. Martin of Alabama Power Company, Carl B. Fritsche of Reinhold Chemicals, and Frank P. Samford of Liberty National Life Insurance Company, all of whom agreed to support financially the mechanics of the scholarship program through the state chamber of commerce. From this cooperation came the General William Crawford Gorgas Scholarship Program, named for Alabama's world-renowned sanitarian.[17]

The first Alabama Science Talent Search was completed in the spring of 1948 and produced eleven white finalists and four black finalists, who had competed separately. Four winners from among the white students each received a four-year tuition scholarship to one of the four participating white colleges. The first-place student chose which of the four institutions he or she wished to attend and received as well a $300 annual allotment toward other fees. The three other finalists chose from the remaining institutions in the order in which they placed and received respectively

$225, $150, and $125 annually. The one black winner received an amount of money equal to that of the white first-place winner, but Tuskegee was the only institution that awarded blacks a tuition scholarship. Of course, these students could choose to attend a college outside the state of Alabama, but the scholarships were not transferable.[18]

The success of this statewide talent search became evident almost immediately. In 1947 only 24 Alabama students entered the national talent search. Following the state awards in the spring of 1948, that fall 153 Alabama students entered the national competition. The 91 white students and 62 black students represented a total of thirty-nine high schools throughout the state. Of the five winners of this first state talent search, all received a bachelor's degree from the institution that they chose to attend, and three of them continued their education in graduate schools. The academy was justifiably proud of its accomplishments.[19]

After the second statewide talent search in the spring of 1949, the state Chamber of Commerce announced that it would no longer be able to raise funds for the Gorgas Scholarships. While it honored its commitment to the ten students from the first two contests, it discontinued awards thereafter. Not willing to give up on the program, however, Thomas W. Martin sought another means of financing the monetary awards. He and Kassner fashioned a tax-exempt foundation to guarantee the future financial security of the awards. Officially organized in 1952, the foundation announced that the contest would resume in 1953 with three additional white colleges and two additional black colleges willing to provide four-year tuition scholarships to the winners. The student response was astounding, as 371 high school seniors entered the competition in 1954. This figure rose to 398 students in 1955 and 546 students in 1956. While some of the winners chose to attend institutions outside the state of Alabama, over half of them remained "at home" for their college education. With only procedural modifications, this scholarship program continues to operate in Alabama today under the auspices of the state academy of science.[20]

Other southern state academies of science cooperated with the talent search in a more limited fashion. In general, they accepted the records from Science Service of students from their states who entered the national competition and chose a number of state winners. Few monetary awards materialized, although the South Carolina academy voted fifty dollars annually to the highest placed South Carolina student. For the most part the state academies encouraged these students to apply for collegiate scholarships and willingly supported their candidacies with letters of reference. No other southern state academy of science has matched Alabama's Gorgas Scholarship Foundation.

Science Fairs

Another program for young people with which southern academies of science became involved was the science fair movement. In 1928 the American Institute of the City of New York launched the science fair program, hoping to stimulate interest in scientific subjects in high schools and encourage the formation of local science clubs. In 1941 Science Clubs of America assumed responsibility for administering this program as well as the National Science Talent Search. Again, southern states were slow to participate. In the early 1950s the Oak Ridge Institute of Nuclear Studies, located in Oak Ridge, Tennessee, decided to promote the movement and hired a field representative, Dewey E. Large, to conduct conferences for teachers, industrialists, and laypersons throughout the South on initiating science fair programs in their states.[21]

Before the end of the decade all of the southern academies of science sponsored state science fairs. Some were conducted in conjunction with junior academy activities, while others operated as entirely separate programs. During the height of the movement, through the 1950s and 1960s, many science teachers required their students to design a suitable project and enter the local competition. Consequently, the science fair movement grew rapidly. By the late 1950s most competitions began at a regional level, where judges chose representatives to participate in the state fair. Winners at the state level then entered national competition. While a number of sponsors of these fairs, including academy members, expressed dismay over the forced participation of students, such a requirement illustrates the widespread concern for developing scientific talent in the United States at this time.[22]

Science fairs attracted many more student participants than did the talent search. In addition to required participation, science fair competition began in junior high school, while the talent search was open only to high school seniors. Moreover, all students regardless of academic standing were eligible to enter a project in the science fair. As always, the success of the state science fair depended on regional funding and a dedicated core of academy members and high school teachers who were willing to invest their energy in the program. Nonetheless, all of the state academies viewed the science fairs as an excellent means to encourage young people to pursue scientific studies, and they were delighted, too, when state winners performed well in the National Science Fair.

Broad Financial Support for Youth Activities

The activities of state academies of science have always been governed by the availability (or inaccessibility) of funds. Generally, both government and private industry have, since the 1940s, contributed more generously to the various youth programs of the academies than to any of their other undertakings. The federal government, following the Soviet Union's shocking success in launching an earth-orbiting satellite in October 1957, suddenly found the wherewithal to invest in America's technological future. The National Science Foundation, created in 1950 by the federal government to furnish grants to graduate students and advanced scholars who exhibited unusual ability, suddenly enjoyed a greatly enlarged budget. NSF administrators, responding to congressional intent, widened the sights of the organization to include a general provision for support of secondary science education at the regional and local levels.[23]

Most of the southern state academies of science hastened to take advantage of this change in policy and applied for NSF grants for a variety of youth-related activities. The Alabama academy's initial request in 1959 for funds to support a full-time director for the betterment of science education in the state was denied, but the organization successfully applied for a grant to fund a visiting scientist program. In 1962 the NSF contributed $8,200 to the Alabama academy's program that offered the services of member scientists to visit high schools, talk to students, and consult with teachers. By 1965 funding had increased to $11,500. During that year the academy sponsored a total of 336 visits to high schools throughout the state. Additional funds from the NSF to the Alabama academy included $10,500 in 1962 to support a one-year study of the effectiveness and potential for junior and collegiate academies and $10,500 in 1965 to finance a program of communication between high school and college teachers.[24]

The North Carolina Academy of Science received twenty thousand dollars from the NSF in 1959 for its proposed short-term institutes for high school teachers. During the academic year 1959–60, the academy sponsored seven such institutes in which high school science teachers received information on the latest scientific developments in their fields, encouragement from conference leaders to sponsor science clubs and otherwise foster the scientific interest of their students, and an invitation to join the academy. By 1962 this program had expanded to eleven institutes and included 291 high school teachers. As with so many other programs begun with funds provided by the NSF, this one declined during the 1960s as NSF monies dried up. By 1970 the North Carolina academy was receiving less than five hundred dollars from the federal organization, and this support was only to meet expenses from former years. The other major project of the North Carolina academy that won NSF approval, a study of geological sites throughout the state that would be considered appropriate for educational field trips, was funded for 1962 and 1963.[25]

The Virginia Academy of Science unsuccessfully applied to the NSF in 1955 for a $2,500 grant to study the results of the Virginia science talent search. However, it won a request in 1959 for $6,450 for a visiting scientist program. Unlike the Alabama academy's visiting scientist program, the Virginia program sponsored a series of two-day visits by academy members to various colleges throughout the state. The academy discontinued this program after two years because of the low federal per diem expense limit. Afterward, it requested NSF funds to support various junior academy activities. Beginning in 1963, the NSF funded the Virginia junior academy annually. The amount fluctuated from a high of $7,770 in 1964 to $2,000 in 1970, and ended in 1971 when the NSF ceased to sponsor junior academies. The Virginia academy also revamped its visiting scientist program along the lines of the one in Alabama and in 1965 received $5,000 from the NSF to sponsor a total of seventy-one visits by academy members to high schools in the state.[26]

The Georgia academy was not so successful in its efforts to obtain grants from the NSF, in part because it did not pursue them with the same vigor as did other academies. Its first proposal, submitted in 1959, requested $39,150 to support a three-year program to improve the junior academy and expand the science fairs. The NSF declined to fund the project. The Georgia academy did receive $8,510 in 1962 to sponsor a series of student visits to college campuses throughout the state, but a similar request for the following year was denied. Evidently this single grant was the only one awarded to the GAS. The academy records make no further mention of proposals to the NSF.[27]

While National Science Foundation monies proved a boon to many of the academies during the 1960s, their decline after 1970 resulted in severe retrenchment of programs that had begun under their auspices. The Virginia academy, for instance, tried to continue the visiting scientist program after the NSF ceased its funding in 1967. During 1968 twenty-seven colleges agreed to meet the expenses of the 143 scientists who volunteered their time to visit high schools; a total of eighty visits occurred in this manner. Such support did not continue, however, and eventually the program met the same fate as the North Carolina academy's teacher institutes.[28] Similar retrenchment was widespread. By 1970 NSF funds to the Alabama academy totaled only $260; to the North Carolina academy, $375. Support of Tennessee academy projects that in 1961 amounted to almost $27,000 had been completely eliminated by this time.[29]

Unfortunately for the academies, the vagaries of federal funding had operated against them. In 1968 Congress passed the Daddario-Kennedy Bill, which authorized the NSF to fund applied as well as basic research and changed five senior staff positions from "career" to "political" status. Although the NSF budget increased markedly as a result of executive attempts to stimulate a sluggish economy, the Office of Management and Budget insisted that the NSF alter many of the programs that it had previously funded. Cut were teacher institutes and other forms of aid to secondary education in favor of increased spending on behalf of small colleges, particularly historically black institutions.[30]

Many academies hoped to continue these programs by appealing to state legislatures for funding. In 1957 Karl Karrison, a member of the Alabama state legislature, had commented that "the Junior Academy would probably have a better chance for money from the state than the Senior Academy."[31] While the Alabama academy never received any state financial support, the Tennessee academy did. Beginning in 1960 the state legislature voted an appropriation of $2,400 to the academy to defray expenses incurred by the science fairs and the junior academy. By 1980 this amount had grown to $15,000, rendering the Tennessee academy the one receiving the most state aid of any in the South. The Tennessee academy divides these funds between various junior academy programs and its Science Teacher Improvement Program, similar in nature to the teacher institutes once sponsored by the North Carolina academy.[32]

Support for the other state academies of science comes not from state governments but from business and industry. Most academies now offer a sustaining membership category for businesses donating one hundred dollars or more annually. In addition, a wide variety of businesses contribute funds to support junior academy activities, particularly the science fairs. Most academy activities that revolve around young people are supported completely by outside donations, leaving the still-modest regular membership fees, usually between ten and twenty dollars annually, to finance the annual meetings and journals.

While youth-oriented activities of the academies are not as expansive as they were during the 1950s and 1960s, neither are they guided by the near hysterical "catch-up" mentality of that era. Currently, science clubs, junior academies, and science fairs direct their programs to students genuinely interested in the sciences. The senior academies hope that local activities might light a spark or develop a talent that will eventually realize its full potential through college, perhaps graduate school, and satisfying employment in some field of scientific endeavor. Consequently, programs for young people remain a high priority for state academies of science.

Looking to the Future

10

If the multidisciplinary state academies of science are anachronisms in this age of specialized, national organizations, as some people have commented from time to time throughout the twentieth century, why have they not collapsed for want of interest and support? Although on shaky ground as they first emerged in the decades prior to World War II, and again in the 1940s, the academies have since rebounded to become effective if not forceful professional organizations. They have survived largely because they recognized and accepted their limitations and because they adapted to the changing needs of the men and women most likely to support them. In this sense, the state academies of science serve as a mirror for the development of professional science in the United States.

Originally, as we have seen, the founders of the academies hoped for thriving societies that would sponsor lively annual meetings, publish creditable journals, and perhaps even provide some incentive and support for research. Such goals reflected the needs of pre–World War II southern scientists, who for the most part were unable to participate in national professional organizations. Unfortunately, these dreams exceeded the financial capacities of the fledgling academies. In the wake of recovery from World War II and the accompanying economic rebirth of the

South, the southern academies expectantly trotted out the old dreams, only to discover that rejuvenated colleges and universities, along with improved communication and transportation, had enabled southern scientists to partake of the benefits of national professional organizations. A reassessment of their situation during the late 1940s and early 1950s produced new objectives for the academies that included continuing the annual meetings, publishing modest journals, and supporting secondary science education through a variety of programs in their respective states. Thus the academies have found a niche for themselves in the world of professional science and expect to continue along the lines that they have developed in the past two decades.

Membership figures attest to the significance of these organizations for many scientists, despite their access to national societies. By 1988 the Georgia academy boasted over 500 members; the Alabama academy reported 878 members. The Virginia academy, one of the largest in the South, fluctuated between 1,400 and 1,650 members during the 1980s. Although large, academy membership is not especially stable. For instance, from March 1988 to March 1989, the Alabama academy dropped 140 members for nonpayment of dues, added 143 new members, and reinstated 28 others. Such a pattern is quite common among state academies of science.[1]

This fluctuating membership should not, though, be misinterpreted as a sign of weakness. Since most academies allow only members to present a paper at the annual meeting and to submit an article for publication, a number of people join for these reasons and then allow their memberships to lapse. Transient members are not necessarily disinterested in the life of the academy; many of them are graduate students or young scholars who soon move out of the state. The academies, since their revitalization after World War II, have in fact recognized that they could provide a first start for young scholars. The academies accept such a role as a part of their contribution to the profession and do not expect everyone who joins the organization to remain a stalwart member.

The number of papers presented each year at annual conventions provides another sign of continued vitality for the state academies. In 1989 the *Journal* of the Alabama Academy of Science published a total of 215 abstracts of presentations from that year's annual meeting. The Tennessee academy's *Journal* in April 1990 contained 82 abstracts from the previous November's meeting. In May 1990 the Georgia academy program listed 113 papers, in addition to a symposium entitled "Science Education for the Twenty-first Century."

Participants continue to represent largely the academic community; less than 10 percent of the presentations came from industrial and governmental circles. Moreover, many of these abstracts list more than one author, indicating professor/graduate student cooperation, which is right in keeping with the modern objective of the academies to assist young scholars. In fact, the Tennessee academy in 1989 voted to provide up to fifty dollars to as many as ten graduate students to defray the expense of presenting a paper at the annual meeting.[2]

Such widespread participation leaves no doubt that the academy meetings are vitally important, and not only to the younger, inexperienced members. More-established scholars who invest much of their energy into training graduate students are pleased that such forums exist and genuinely enjoy seeing their efforts, and those of their colleagues, "pay off" in academic terms. Then too, senior scientists occasionally use this forum themselves, to foretell of research that they may report in more detailed fashion elsewhere. The more intimate academy meetings, as opposed to those of national societies that attract thousands of participants, provide a friendly audience for feedback. These annual meetings are the single most important activity of the state academies and show no signs of declining in the future.

The second most important general focus for modern state academies of science, and their most effective role as public servant, relates to their desire to stimulate scientific awareness among the region's young people. Although state academies since their inception have expressed concern over this issue, they realized during the post–World War II reorganization era that their position as (relatively) small, statewide organizations afforded them a unique opportunity to address the matter of education. First, they could keep abreast of state educational policies and communicate with government officials who must remain alert to the needs of their constituency. Second, as Arthur Bevan of the Illinois academy noted in 1951, state academies stood a better chance than national scientific societies of attracting members from among high school and small college teachers, as well as from graduate and under-graduate students.[3] Through programs designed to address the needs of these persons, state academies have exerted a significant effect on the quality of science education within the state.

The academies approach this objective in different manners. A number of academies have inaugurated a science education section; papers at each year's annual meeting are devoted to improving techniques in the classroom and other methods for stimulating a love of scientific inquiry. All of the state academies of science sponsor some sort of junior academy activities. Some of the junior organizations meet in conjunction with the senior academies; others convene in regional meetings around the state; still others hold one annual convention. Science fairs also remain an integral component of contact between senior academies and high school students.

The Alabama academy, which was the first one in the South to sponsor a junior academy, has continued to sponsor a strong program for young scholars. In addition to junior academy meetings, the Alabama academy supports annual science fairs and coordinates the Gorgas Scholarship competition, begun in 1947. In 1989 twenty finalists were chosen from regional science fairs to represent Alabama at the national science fair. The Gorgas Scholarship competition in 1989 selected three winners, who received $2,500, $1,500, and $1,000 tuition grants, as well as ten additional finalists. All of the students received scholarship offers from one of Alabama's institutions of higher education. Funding for this program relies not on yearly contributions or legislative grants but is solidly held in trust, boding well for the future.[4]

The Tennessee academy likewise has a strong program for the state's young people. It maintains a visiting scientist program offered to state schools at no cost, whereby academy members agree to visit schools that make a request for such a visit. In the 1988–89 academic year, 170 scientists agreed to participate in the program;

sixty-eight schools requested visits, thirty were completed, and sixteen were still pending at the time the proceedings were published. The academy estimated that it reached approximately 2,900 students and 100 teachers in this manner. Such an expensive program is not sustained easily. Fortunately for the academy, the state legislature contributes annually to its maintenance; in 1989 the state appropriated nineteen thousand dollars for academy coffers. An end to this funding would no doubt spell disaster for the Visiting Scientist Program.[5]

The Tennessee academy also sponsors both a junior and a collegiate academy. The former meets annually, with students from around the state competing for the honor of presenting a paper at the meeting. In 1989 a total of twenty-three presentations were invited from among forty-four submissions. Additionally, two students received partial funding to attend the American Junior Academy of Science, which met jointly with the AAAS. The collegiate academy is somewhat more loosely structured and holds three regional meetings each year.[6]

In addition to statewide meetings, a few of the academies offer modest research awards (not scholarships) to high school and collegiate scholars. During the 1989–90 academic year, for instance, the Tennessee academy distributed $1,900 among twenty-four students from nine schools who had collectively requested $6,300. The AAAS provided $900 of this money, a vestige of the old research award first implemented in the 1930s; the remainder came from academy funds.[7]

In part this move to award research grants to students rather than to senior academy members stems from the late 1950s, when the United States was fearful that it was behind the Soviet Union in science and technology and must catch up. At that time the AAAS required that its contribution to state academies be utilized in this manner; National Science Foundation grants also stipulated that their money be spent on educational programs.

Additionally, in this age of large research grants, the modest sums available to state academy research committees simply are not large enough to make applying for them worthwhile for academic scientists. Even the Virginia academy, which exerted a major effort under the direction of Shelton Horsley to build a research endowment, entered the decade of the 1980s with a trust fund of only forty-three thousand dollars. Rather than discontinue research grants completely, the academies have almost universally chosen to utilize their modest funds to assist secondary, collegiate, and graduate students, for whom a few hundred dollars can make a significant difference. This alteration of policy reflects the academies understanding of their changed role in the life of professional science and their willingness to find and fill a productive niche, albeit different from the one originally conceived so many decades ago.

Some programs begun by state academies of science have not survived the test of time as well as junior academy activities. One example is the demise of the Tennessee academy's Reelfoot Lake Biological Research Station, begun in 1932. In 1977 Wintfred L. Smith, director of the program, announced that "the active life of the present Reelfoot Lake Biological Station has effectively ended. Time and vandalism have taken their toll on the building and in its current state it is considered unsafe for use." Smith and other members of the advisory committee recommended that the academy relinquish its lease on the property. Noting that over two hundred publications had resulted from study of the Reelfoot Lake area, Smith suggested that perhaps the academy could "pursue the establishment of a modest station at a different lake site." However, economic realities made this a dim prospect.[8]

In fact, most state academies of science operate with rather austere budgets. The total income for the Georgia academy in 1989 was $16,750; $7,500 came from dues, $3,450 represented the University of Georgia's subsidy for publication of the journal, $2,000 came from contributions, $1,600 from meeting registration, and the remainder from journal subscriptions, exhibit fees, and interest. With anticipated expenses of $14,000 for publication of the journal alone, few funds remain for other academy activities.[9] The Tennessee academy fares much better, thanks to state support, with total 1989 receipts amounting to $32,604. Should the legislature reduce or terminate its support, the Tennessee academy would find itself in dire straits indeed; undoubtedly junior academy work, which amassed expenses in excess of $8,000 in 1989, would be in jeopardy.[10]

Given their modest financial base, state academies of science show no promise of expanding their current programs. They do, however, exhibit considerable stability within the bounds proscribed by financial strictures. First, membership remains strong, as does attendance at the annual meetings. While rosters indicate a 10 to 12 percent turnover each year, such a figure is not alarming simply because it indicates that the academies are doing one of their jobs—offering a start in the profession to young scholars.

Second, the publications of the academies no longer attempt to compete with those of national organizations but operate within their financial abilities. Most of them publish abstracts from the annual meeting, which afford scientists an opportunity to alert the professional world to research in progress. The few articles that appear annually again provide a starting point for young scholars. Finally, reports of academy activities keep members abreast of the junior academy and any other programs that the organization might sponsor.

Third, the various programs designed to stimulate scientific education and inquiry among the states' youth afford member scientists the opportunity to make a difference for hundreds of students and provide an invaluable service for the students themselves. Such programs as junior academies, science fairs, and visiting scientists are not likely to be sponsored by any other organization, let alone state governments. The academies have thus found a purpose that does not require competition with national organizations and that is sometimes able to attract outside funding.

By altering their focus, especially during the 1940s and 1950s, state academies of science discovered a means of remaining a significant if not a vital force within the lives of professional scientists. They are an important component of the career ladder, especially for young scholars, and provide an opportunity for community service to all members. Although presentations at academy conferences do not carry the prestige of those at national meetings, scientists at all stages of their career attend and participate. State meetings provide a less formal and more friendly atmosphere away from the pressure of the daily world for scientists to discuss their work with congenial and informed colleagues. With national organizations now consisting of thousands of members, most of whom do not know each other even by reputation, more than one scientist has indicated that the more casual atmosphere of the state academy assemblies, if no other reason, brings him or her back each year.

In the final analysis, the academies will survive and prosper, within their limitations, as long as they continue to meet the professional needs of a significant number of scientists. They were born of those professional needs, and as the needs of regional scientists changed, so did the academies. Given the fast pace of the modern world, it is impossible to predict the paths that scientific careers will take and the demands that scientists will place upon their professional organizations. Nonetheless, the academies must remain alert, as they have in the past, to alterations that would affect their vitality and be prepared to respond accordingly. Their key to survival lies in their own history.

→ A great prospective prescriptive

Notes

Abbreviations

AAAS	American Association for the Advancement of Science
AAS	Alabama Academy of Science
AISS	Alabama Industrial and Scientific Society
BGAS	*Bulletin of the Georgia Academy of Science*
BGSA	*Bulletin of the Geological Society of America*
BSCAS	*Bulletin of the South Carolina Academy of Science*
EMSS	Elisha Mitchell Scientific Society
GAS	Georgia Academy of Science
GJS	*Georgia Journal of Science*
JAAS	*Journal of the Alabama Academy of Science*
JEMSS	*Journal of the Elisha Mitchell Scientific Society*
JMAS	*Journal of the Mississippi Academy of Science*
JTAS	*Journal of the Tennessee Academy of Science*
NAS	National Academy of Science
NCAS	North Carolina Academy of Science
NSF	National Science Foundation
PIAS	*Proceedings of the Indiana Academy of Science*
PLAS	*Proceedings of the Louisiana Academy of Science*
PVAS	*Proceedings of the Virginia Academy of Science*
SASI	Southern Association for Science and Industry
SCAS	South Carolina Academy of Science
TAS	Tennessee Academy of Science
TKAS	*Transactions of the Kansas Academy of Science*
TTAS	*Transactions of the Tennessee Academy of Science*
TVA	Tennessee Valley Authority
UNC	University of North Carolina
VAS	Virginia Academy of Science
VJS	*Virginia Journal of Science*

1. Organization: A Vital Component of Professionalism

1. Thomas L. Haskell, *The Emergence of Professional Social Science: The American Social Science Association and the Nineteenth-Century Crisis of Authority* (Urbana: University of Illinois Press, 1977), 1.

2. See, for instance, George Daniels, "The Process of Professionalization in American Science: The Emergent Period, 1820–1960," *Isis* 58 (1967): 151–66; Robert Bruce, "A Statistical Profile of American Scientists, 1846–1876," in *Nineteenth-Century American Science: A Reappraisal*, ed. George Daniels (Evanston, Ill.: Northwestern University Press, 1972), 63–94; Bruce, *The Launching of Modern American Science, 1846–1876* (New York: Alfred A. Knopf, 1987); Nathan Reingold, "Definitions and Speculations: The Professionalization of Science in America in the Nineteenth Century," in *The Pursuit of Knowledge in the Early American Republic: American Scientific and Learned Societies from Colonial Times to the Civil War*, eds. Alexandra Oleson and Sanborn C. Brown (Baltimore: Johns Hopkins University Press, 1976), 33–69; John J. Beer and W. David Lewis, "Aspects of the Professionalization of Science," *Daedalus* 92 (1963): 754–84.

3. Reingold, "Definitions and Speculations."

4. Ibid., 39–41.

5. Sally Gregory Kohlstedt, *The Formation of the American Scientific Community: The American Association for the Advancement of Science, 1848–60* (Urbana: University of Illinois Press, 1976), 59–77.

6. Ibid., 79.

7. Ibid., 66–79.

8. Ibid., 228–29; A. Hunter Dupree, *Science in the Federal Government: A History of Policies and Activities to 1940* (Cambridge: Harvard University Press, Belknap Press, 1957), 135–41, 300.

9. Daniel J. Kevles, *The Physicists: The History of a Scientific Community in Modern America* (1971; reprint, Cambridge: Harvard University Press, 1987), 40.

10. Henry A. Rowland, "A Plea for Pure Science," AAAS *Proceedings* 32 (1883): 113.

11. Ibid., 110–11.

12. Kevles, Daniel J., Jeffrey L. Sturchio, and P. Thomas Carroll, "The Sciences in America, circa 1800," *Science*, July 5, 1980, pp. 27–32.

13. Kevles, *Physicists*, 35–36.

14. Burton J. Bledstein, *The Culture of Professionalism: The Middle Class and the Development of Higher Education in America* (New York: W. W. Norton, 1976), x.

15. Clement Eaton, *The Mind of the Old South*, rev. ed. (Baton Rouge: Louisiana State University Press, 1967), 243; more recent scholarship includes Ronald L. Numbers and Janet S. Numbers, "Science in the Old South: A Reappraisal," *Journal of Southern History* 48 (May 1982): 163–84; Lester D. Stephens, *Joseph LeConte: Gentle Prophet of Evolution* (Baton Rouge: Louisiana State University Press, 1983); and Tamara Miner Haygood, *Henry William*

Ravenel, 1814–1887: South Carolina Scientist in the Civil War Era (University: University of Alabama Press, 1987).

2. Scientists in the Postbellum South

1. Stephens, *Joseph LeConte*, 89–91.
2. Southern states included in this percentage are Kentucky, Tennessee, Missouri, Virginia, North Carolina, South Carolina, Georgia, Florida, Alabama, Louisiana, Mississippi, and Arkansas. See Bruce, "Statistical Profile," 79.
3. Stephens, *Joseph LeConte*, 104–12.
4. Kemp P. Battle, *History of the University of North Carolina*, 2 vols. (Raleigh: Edwards & Broughton, 1907–12; reprint, Spartanburg, S.C.: Reprint Company, 1974), 1:784–85.
5. James B. Sellers, *History of the University of Alabama*, vol. 1, *1818–1902* (University: University of Alabama Press, 1953), 285, 291–92.
6. Thomas G. Dyer, *The University of Georgia: A Bicentennial History, 1785–1985* (Athens: University of Georgia Press, 1985), 115–17.
7. University of North Carolina *Alumni Quarterly* 1 (October 1894): 33–34.
8. Virginius Dabney, *Mr. Jefferson's University: A History* (Charlottesville: University Press of Virginia, 1981), 28; Philip Alexander Bruce, *History of the University of Virginia, 1819–1919*, 5 vols. (New York: Macmillan, 1921), 3:355, 359–61.
9. Robert C. McMath, Jr., Ronald H. Bayor, James E. Brittain, Lawrence Foster, August W. Giebelhaus, and Germaine M. Reed, *Engineering in the New South: Georgia Tech, 1885–1985* (Athens: University of Georgia Press, 1985), 9.
10. Charles Butts, "Memorial of Eugene Allen Smith," *BGSA* 39 (1927): 51–52; "Death of Mrs. Julia Allen," newspaper clipping (source not noted) in Scrapbook 5, p. 69. Smith avidly clipped items of interest to him and pasted them in notebooks, now located in the Amelia Gayle Gorgas Library, University of Alabama, Tuscaloosa.
11. Butts, "Memorial of Eugene Allen Smith," 51–52; Walter B. Jones, "Eugene Allen Smith," *Science*, January 6, 1928, p. 7; Sellers, *History of the University of Alabama*, 279.
12. Butts, "Memorial of Eugene Allen Smith," 52; Eugene Allen Smith, "Memorial of Eugene Woldemar Hilgard," *BGSA* 28 (1916): 41–42.
13. Eugene Allen Smith to W. H. Wahl, April 23, 1873, in Stewart J. Lloyd, *Eugene Allen Smith: Alabama's Great Geologist* (New York: Newcomen Society in North America, 1954), 22.
14. Butts, "Memorial of Eugene Allen Smith," 59–65. The *BGSA* article lists all of Smith's publications.
15. Maurice M. Bursey, *Carolina Chemists: Sketches from Chapel Hill* (Chapel Hill: University of North Carolina Press, 1982), 35.
16. Ibid.
17. Ibid., 36–37.

18. Ibid., 37–39.
19. Ibid., 47, 50–51, 64–66.
20. Battle, *History of the University of North Carolina*, 2:257; William S. Powell, ed., *Dictionary of North Carolina Biography*, 2 vols. (Chapel Hill: University of North Carolina Press, 1986), 2:320–21.
21. Joseph Hyde Pratt, "Memorial Sketch of Dr. Joseph Austin Holmes," *JEMSS* 32 (April 1916): 1–15; Francis Preston Venable, "Joseph Austin Holmes," ibid., 16–19; Kemp P. Battle, "Dr. Joseph Austin Holmes at the University of North Carolina," ibid., 20–23.
22. Venable, "Joseph Austin Holmes," 17.
23. Ibid., 16–17.
24. Francis Preston Venable, "Historical Sketch of the Elisha Mitchell Scientific Society," *JEMSS* 39 (April 1924): 119; Battle, "Dr. Joseph Austin Holmes," 20.
25. Venable, "Joseph Austin Holmes," 17.
26. Venable, "Historical Sketch," 119.
27. Ibid., 117–18; Archibald Henderson, "The Elisha Mitchell Scientific Society: Its History and Achievements," *JEMSS* 50 (December 1934): 2; Francis Preston Venable, "The Elisha Mitchell Scientific Society—Historical Sketch, Growth and Development," *Carolina Chemist* 8 (May 1922): 10–11; Venable, "The Mitchell Society," *Alumni Quarterly* 1 (October 1894): 39–41.
28. A copy of the original circular is contained in the records of the EMSS, in the possession of Maurice Bursey, Department of Chemistry, University of North Carolina, Chapel Hill.
29. Ibid.
30. Venable, "Historical Sketch," 117.
31. Constitution of the EMSS, *JEMSS* 1 (1884): 90–92.
32. President's Report for 1884, *JEMSS* 1 (1884): 3.
33. Constitution of the EMSS, *JEMSS* 1 (1884): 90–92.
34. Ibid., 91.
35. Francis Preston Venable, "Inaugural Address before the Mitchell Scientific Society," November 10, 1883, ms. copy in EMSS Records, 5–6.
36. Ibid., 6–8.
37. Ibid., 7.
38. President's Report for 1884, *JEMSS* 1 (1884): 4; list of papers presented before the society, ibid., 6–8.
39. Treasurer's Report, *JEMSS* 1 (1884): 8; Membership List, ibid., 93–94.
40. Report of the Secretary, *JEMSS* 1 (1884): 5–6.
41. Programs of meeting, *JEMSS* 2 (1885): 3, 6–7; report of the Resident Vice-President for the Year 1884–85, ibid., 3.
42. Venable, "Mitchell Society," 39–41.
43. Battle, *History of the University of North Carolina*, 2:512.
44. List of Exchange Societies, *JEMSS* 5 (1888): 50–53.
45. "Little Doc's Ambulance," pamphlet issued by the University of Alabama in commemoration of the restoration of Smith's Studebaker wagon, n.d.
46. Ethel Armes, *The Story of Coal and Iron in Alabama* (Birmingham, Ala.: Chamber of Commerce, 1910; reprint, New York: Arno Press, 1973), 265.
47. *Proceedings of the Alabama Industrial and Scientific Society* 1 (1891): 3–4.
48. Ibid., 3–7.

49. Ibid., 8–9.
50. Ibid.
51. Ibid., 72; and 3 (1893): 6.
52. *AISS Proceedings* 9 (1899): 39.
53. *Birmingham News*, March 15, 1900.

3. The Need for Organization: Emergence of State Academies of Science

1. Ralph S. Bates, *Scientific Societies in the United States*, 3d ed. (Cambridge: MIT Press, 1965), 14–15.
2. John Hendley Barnhart, "The First Hundred Years of the New York Academy of Sciences," *Scientific Monthly* 5 (November 1917): 463–75.
3. Bates, *Scientific Societies*, 47–48; William M. Bailey, "The Beginning of the Illinois State Academy of Science," *Illinois Academy of Science Transactions* 43 (1950): 24; Amos W. Butler, "Early History of the Indiana Academy of Science," *PIAS* 33 (1923): 14–15.
4. W. A. Harshbarger, "The Kansas Academy of Science," *TKAS* 29 (1918): 35–41; W. H. Schoewe, "The Kansas Academy of Science—Past, Present, and Future," ibid. 41 (1938): 399–413.
5. See references in n. 4.
6. Currently the California Academy of Science is the largest and best-endowed academy in the nation, with eleven thousand members and an annual budget of approximately $3 million. See Bates, *Scientific Societies*, 48; Joseph Ewan and Nesta Ewan, "San Francisco as a Mecca for Nineteenth Century Naturalists," in *A Century of Progress in the Natural Sciences, 1853–1953* (San Francisco: California Academy of Sciences, 1955), 9–33 (LeConte quotation on p. 32); *Directory and Handbook of the Association of Academies of Science, 1978–1979* (Columbus, Ohio, 1979), p. 15.
7. Herbert Hutchinson Brimley to W. C. Coker, n.d., William Willard Ashe Papers, Southern Historical Collection, University of North Carolina, Chapel Hill.
8. J. S. Holmes, untitled article, n.d., Ashe Papers.
9. Service Bulletin, United States Forest Service, March 28, 1932, Ashe Papers.
10. Eugene Pleasants Odum, "Introduction," in *A North Carolina Naturalist, H. H. Brimley: Selections from His Writings*, ed. Odum (Chapel Hill: University of North Carolina Press, 1949).
11. H. H. Brimley, "The Founding of the N.C. Academy of Science," undated manuscript in the records of the North Carolina Academy of Science, Special Collections, J. Y. Joyner Library, East Carolina University, Greenville, North Carolina.
12. H. H. Brimley to William Louis Poteat, March 8, 1902, Herbert Hutchinson Brimley Papers, North Carolina State Archives, Raleigh.
13. H. H. Brimley to T. Gilbert Pearson, March 8, 1902, NCAS Records.
14. Brimley, "Founding of the N.C. Academy of Science."

15. Ibid.
16. Ibid.
17. Ibid.
18. Ibid.; NCAS Minute Book, NCAS Records; list of First Fellows of the NCAS, Brimley Papers.
19. See Frank J. Jeter, "Benjamin Wesley Kilgore," and Clarence Poe, "Benjamin Wesley Kilgore—His Life and Work," both printed in the program for the dedication of Kilgore Hall at North Carolina State University, December 18, 1953, Biographical Clippings File, North Carolina State University Archives, D. H. Hill Library, Raleigh. An examination of Kilgore's personal correspondence for 1902 reveals no mention to anyone of this organization; evidently it meant little to him. See Benjamin Wesley Kilgore Papers, North Carolina State Archives, Raleigh.
20. *Raleigh News and Observer*, March 23, 1902, p. 4.
21. Brimley, "Founding of the N.C. Academy of Science"; minutes of the meeting of the NCAS, November 28–29, 1902, *JEMSS* 20 (1904): 3–4; *Raleigh News and Observer*, November 28, 1902, p. 7.
22. Minutes of the meeting of the NCAS executive committee, May 1, 1903, *JEMSS* 20 (1904): 5–6.
23. Ibid.; Constitution of the NCAS, ibid., 17.
24. Proceedings of the NCAS, *JEMSS* 21 (1905): 57; *Raleigh News and Observer*, May 10, 1905, p. 5.
25. Proceedings of the NCAS, *JEMSS* 25 (1909): 39–42; proceedings of the NCAS, ibid. 26 (1910): 51–53.
26. B. F. D. Runk, "Ivey Foreman Lewis," *VJS* 5 (April 1954): 51–52.
27. Proceedings of the NCAS, *JEMSS* 26 (1910): 51.
28. Charles Baskerville, "Science and the People," *JEMSS* 20 (1904): 75.
29. Ibid., 78.
30. Ibid.
31. Proceedings of the NCAS, *JEMSS* 25 (1909): 41.
32. Proceedings of the NCAS, *JEMSS* 26 (1910): 51–52.
33. "Report of the Committee of the North Carolina Academy of Science on Science Teaching in North Carolina," *North Carolina High School Bulletin*, July 1911, reprint in NCAS Records.
34. Ibid.; W. C. Coker, "Science Teaching," *North Carolina High School Bulletin*, April 1910, reprint in NCAS Records.
35. Collier Cobb, "The Garden, Field, and Forest of the Nation," *JEMSS* 23 (1907): 52–70, quotations on pp. 69 and 54.
36. Joseph Hyde Pratt, "The Conservation and Utilization of Our Natural Resources," *JEMSS* 26 (1910): 1–25, quotation on p. 4.
37. George Martin Hall, "Memorial of Charles Henry Gordon, First President of the Tennessee Academy of Science," *JTAS* 10 (1935): 101–3.
38. The letter sent by Gordon to his colleagues was reprinted in *TTAS* 1 (1914): 5–6.
39. Proceedings of the TAS, *TTAS* 1 (1914): 6–7.
40. Constitution of the TAS, *TTAS* 1 (1914): 8–10.
41. Proceedings of the TAS, *TTAS* 1 (1914): 6–7.
42. Ibid., 18–20.

43. Charles Henry Gordon, "Science and Progress in the South," *TTAS* 1 (1914): 52.

44. Ibid., 54–56.

45. Report of the meeting of the TAS, November 29–30, 1912, *TTAS* 1 (1914): 18.

46. Ibid., 18–19.

47. Report of the meeting of the TAS, April 10, 1914, *TTAS* 2 (1917): 9.

48. J. T. McGill, "The Tennessee Academy of Science," *JTAS* 2 (1927): 7–10.

49. *Proceedings of the Virginia Academy of Science,* 1923–24, pp. 4–5; Harry Joseph Staggers, "A History of the Virginia Academy of Science, 1923–1945," *VJS* (Winter 1968): 61–62.

50. A copy of the letter of February 14, 1922, is contained in the records of the Georgia Academy of Science, currently in the possession of George A. Rogers, Georgia Southern College, Statesboro.

51. Ibid.; R. P. Stephens, "The Georgia Academy of Science, 1922–1942," undated manuscript in GAS Records, 1–2.

52. B. B. Higgins to T. H. McHatton, March 13, 1922, GAS Records.

53. Minutes of the executive council meeting, February 3 and October 27, 1928, and minutes of the meeting of the GAS, April 3–4, 1936, GAS Records.

54. W. S. Nelms, Presidential Address, n.d. [1924], manuscript copy in GAS Records.

55. W. A. Gardner to S. A. Ives, June 29, 1923, records of the Alabama Academy of Science, Archives, Draughon Library, Auburn University, Auburn, Alabama, Ellen B. Buckner, "Outstanding Alabama Scientists: Profiles of Five Years' Recipients of the Wright A. Gardner Award," *JAAS* 60 (April 1989): 49.

56. Eugene Allen Smith to W. A. Gardner, January 21, 1924, AAS Records; proceedings of the AAS, April 4, 1924, *JAAS* 4 (1933): 38–39. See also Wright A. Gardner, "The Organization of the Alabama Academy of Science," ibid. 3 (1932): 7–9.

57. Notice of meeting of April 4, 1924, *Science,* May 30, 1924, p. 481; Earl E. Sechriest to W. A. Gardner, January 14, 1924, and Gardner to Sechriest, January 21, 1924, AAS Records.

58. W. E. Hoy, Jr., "History of the South Carolina Academy of Science," *BSCAS* 1 (1935): 2–3; Constitution of the SCAS, ibid. 3 (1937): 20–22; proceedings of the SCAS, March 26, 1925, minutebook, records of the South Carolina Academy of Science, South Caroliniana Library, University of South Carolina, Columbia; Harry J. Bennett, "The First Half-Century of the Louisiana Academy of Science, 1927–1977," *PLAS* 45 (December 1982): 9–13.

59. L. B. Ham, "Early History: The Arkansas Academy of Science," *Proceedings of the Arkansas Academy of Science* 1 (1941): 3–6.

60. J. H. Kusner, "The Academy during 1936," *Proceedings of the Florida Academy of Sciences* 1 (1937): 1–2; E. Ruffin Jones, Jr., "The Evolution of the Academy of Science," *Quarterly Journal of the Florida Academy of Sciences* 24 (December 1961): 229–30.

61. Gordon Gunter, "The Atmosphere Surrounding Science in Mississippi,"

JMAS 12 (1966): 29–44; Ernest E. Russell, "Status of the Academy of Science Today," ibid. 14 (1968): vii–x; editor's preface, ibid. 11 (1965): iii.

4. Support for Fledgling Academies:
North Carolina Scientists, a Case Study

1. David D. Whitney, "State Academies of Science," *Science*, December 5, 1919, pp. 517–18.
2. Ibid.
3. T. C. Mendenhall, "The Relation of the Academy to the State and to the People of the State," *Science*, December 24, 1915, pp. 881–90.
4. Ibid.
5. Paul P. Boyd, "The Future of the State Academy of Science," *Science*, June 11, 1920, pp. 575–80.
6. W. S. Bayley, "The Place of State Academies of Science among Scientific Organizations," *Science*, June 1, 1923, pp. 623–29.
7. Ibid., 629.
8. Several recent studies have added significantly to an understanding of the development of southern higher education. See especially Dyer, *University of Georgia,* and Paul Conkin, *Gone with the Ivy: A Biography of Vanderbilt University* (Knoxville: University of Tennessee Press, 1985).
9. Annual catalogs, University of North Carolina, North Carolina State College, Trinity College, Wake Forest College, Davidson College, and Elon College.
10. The information in the text and in the tables concerning these 498 North Carolina scientists was obtained from college catalogs for the eight institutions included in the survey for the time period 1875–1940.
11. J. McKeen Cattell, ed. *American Men of Science* (Lancaster, Pa.: Science Press, 1906). Editions used to accumulate biographical information for this chapter are the 4th (1927), 5th (1933), 6th (1938), and 7th (1944). All information on individuals is based on this source unless specifically noted otherwise.
12. Bursey, *Carolina Chemists,* 124.
13. Cattell, *American Men of Science.* The introductions to the first four editions (1906, 1910, 1921, and 1927) outline the procedure whereby the outstanding scientists were selected.
14. Hugh T. Lefler and Albert Ray Newsome, *North Carolina: The History of a Southern State,* 3d rev. ed. (Chapel Hill: University of North Carolina Press, 1973), 556, 559.
15. A. C. Howell, *The Kenan Professorships* (Chapel Hill: University of North Carolina Press, 1956), 3–72.
16. Ibid., 172–73.
17. Ibid., 185–89.
18. Bursey, *Carolina Chemists,* 87–88.
19. Howell, *Kenan Professorships,* 109–13; Archibald Henderson, *The Campus of the First State University* (Chapel Hill: University of North Carolina Press, 1949), 234–40.

20. Howell, *Kenan Professorships*, 209–11.
21. Ibid., 217–19.

5. Camaraderie, Research, and Publication:
State Academies and Scientists' Professional Needs

1. All figures were obtained from secretaries' reports printed in the academy journals.
2. Ibid. Secretary reports usually noted at least the total number of papers presented; in some instances each author and title was listed. The number of papers presented before the Georgia academy remained somewhat static, averaging about twenty-seven, because of the limited membership. The Louisiana academy's printed proceedings contain no record of papers presented.
3. C. L. Baker, *History of the Academy Conference, 1926–1952* (Washington, D.C.: American Association for the Advancement of Science, 1952), 2–16; report of the organization of the Academy Conference, *Science*, November 30, 1928, p. 533.
4. McGill, "Tennessee Academy of Science," 8.
5. Ibid., 7–10; proceedings of the TAS, *TTAS* 1 (1914) and 2 (1917); editorial, *JTAS* 5 (January 1930): 25–26; meeting of executive committee of TAS, April 27, 1931, ibid. 6 (July 1931): 154.
6. Bayley, "Place of State Academies," 623–29; Boyd, "Future of the State Academy of Science," 575–80.
7. Francis Preston Venable, "Isotopes," *JEMSS* 37 (1921–22): 122.
8. Proceedings of the AAS, March 26–27, 1926, *JAAS* 4 (1933): 93; proceedings of the AAS, March 30–31, 1928, ibid. 2 (1930): 41; Clyde H. Cantrell, Paul C. Bailey, and S. B. Barker, eds., *A History of the Alabama Academy of Science* (Auburn: Alabama Academy of Science, 1963), 48.
9. Minutes of the SCAS executive council meeting of April 10, 1940, *BSCAS* 6 (1940): 3–5.
10. Bennett, "First Half-Century," 41–42.
11. Allan D. Charles, "A History of the Georgia Academy of Science" (Master's thesis, Emory University, 1968), 118–20; minutes of the GAS executive committee meeting, December 1, 1922, and February 21, 1930, GAS Records.
12. *PVAS*, 1929–30, p. 5.
13. *PVAS*, 1933–34, p. 13; Staggers, "History of the Virginia Academy of Science," 76; Harry Joseph Staggers and Walter S. Flory, "A History of the Virginia Academy of Science, 1945–1973," *VJS* 24 (Winter 1973): 11.
14. *TTAS* 1 (1914): 8; Constitution of the AAS, in Cantrell, Bailey, and Barker, *History of the Alabama Academy of Science*, 105.
15. Staggers, "History of the Virginia Academy of Science," 65, 71; President's Report, *PVAS*, 1933–34, p. 7.
16. Proceedings of the NCAS, May 4–5, 1934, *JEMSS* 50 (December 1934): 26;

proceedings of the AAS, April 14, 1944, *JAAS* 16 (September 1944): 10; report of the GAS secretary, *BGAS* 4 (July 1946): 1.

17. Patrick Yancey to Russell S. Poor, April 13, 1936, AAS Records.
18. Scott C. Lyon, "Possibilities for a Biological Station at Reelfoot Lake," *JTAS* 1 (January 1926): 10–15.
19. Ibid., 12–15.
20. J. T. McGill, "History of the Reelfoot Lake Biological Station," *JTAS* 8 (January 1933): 23–24.
21. Ibid., 26–32.
22. Report of the Director of the Reelfoot Lake Biological Station and report of the Trustees of the Tennessee Academy of Science, *JTAS* 14 (January 1939): 2–5.
23. J. McKeen Cattell, "Local Branches of the American Association for the Advancement of Science," *Science*, December 21, 1934, p. 576.
24. Ibid.; Burton E. Livingston to Septima Smith, July 8, 1935, and A. G. Overton to Patrick Yancey, October 24, 1935, AAS Records.
25. A. G. Overton to Patrick Yancey, October 24, 1935, and Overton to B. F. Clark, January 21, 1936, AAS Records.
26. Meeting of the SCAS executive council, April 24, 1936, *BSCAS* 2 (1936): 2–3; meeting of the TAS executive committee, November 14, 1935, *JTAS* 11 (January 1936): 68. The discussion of the matter among the members of the NCAS is contained in the following correspondence: Milton L. Braun to R. E. Coker, October 21, 1935; Coker to C. F. Korstian, October 26, 1935; Korstian to Coker, October 31, 1935; R. F. Poole to Coker, November 1, 1935; Coker to H. L. Blomquist, November 4, 1935, Robert Ervin Coker Papers, Southern Historical Collection, University of North Carolina, Chapel Hill.
27. C. F. Korstian to R. E. Coker, November 29, 1935; Milton Braun to Coker, November 29, 1935, Robert Ervin Coker Papers; proceedings of the NCAS, April 24–25, 1936, *JEMSS* 52 (December 1936): 139.
28. O. J. Thies to W. C. Coker, June 28, 1937, William C. Coker Papers, Southern Historical Collection, University of North Carolina, Chapel Hill.
29. Proceedings of the SCAS, April 25, 1936, *BSCAS* 2 (1936): 6; minutes of SCAS council meeting, April 30, 1937, ibid. 3 (1937): 4–5; report of the SCAS research committee, ibid. 4 (1938): 5; report of the SCAS research committee, ibid. 5 (1939): 5.
30. A. G. Overton to Patrick Yancey, October 24, 1935, AAS Records; proceedings of the AAS, March 20–21, 1936, *JAAS* 8 (1936): 7; proceedings of the AAS, April 14–15, 1939, ibid. 11 (June 1939): 6; proceedings of the AAS, March 29–30, 1940, ibid. 12 (June 1940): 8.
31. Reports of the TAS research committee in the following issues of the *JTAS*: 12 (1937): 228; 13 (1938): 57; 14 (1939): 220; 15 (1940): 325; 16 (1941): 249.
32. Bennett, "First Half-Century," 11–12, 50–51.
33. Staggers, "History of the Virginia Academy of Science," 65; J. Shelton Horsley to William M. Brown, December 21, 1926, records of the Virginia Academy of Science, Special Collections, Virginia Polytechnic Institute and State University, Blacksburg.
34. E. C. L. Miller to R. F. McCrackan, June 7, 1926; report of the Endowment Fund, March 4, 1927, VAS Records.

35. Joseph C. Aub to J. Shelton Horsley, March 1, 1935, VAS Records; reports of the research committee, *PVAS*, consecutive issues from 1929–30 to 1938–39; report of the research committee, *VJS* 1 (November 1940).

36. E. C. L. Miller to J. Shelton Horsley, November 10, 1931, VAS Records.

37. The annual reports of the VAS research committee, VAS Records.

38. J. Shelton Horsley to W. T. Sanger, May 18, 1937; Horsley to E. C. L. Miller, May 29, 1937, VAS Records.

39. A copy of Horsley's master letter is located in the records of the endowment fund for 1937, VAS Records.

40. AAS secretary's report, *JAAS* 12 (June 1940): 12–13.

41. Lloyd C. Bird to George H. Boyd, July 29, 1935, GAS Records; Staggers, "History of the Virginia Academy of Science," 72–93; President's Report, *PVAS*, 1935–36, pp. 9–10; proceedings of the NCAS, May 7–8, 1937, *JEMSS* 53 (December 1937): 209, and April 30–May 1, 1943, ibid. 59 (December 1943): 96; SCAS council meeting, April 24, 1935, *BSCAS* 2 (1936): 2–3.

42. Gillie A. Larew to E. C. L. Miller, January 23, 1943, VAS Records, Miller to E. V. Jones, November 4, 1943, AAS Records.

43. Otis Caldwell to E. V. Jones, January 11, 1944, and C. M. Goethe to Jones, April 3, 1944, AAS Records.

44. Otis Caldwell to E. V. Jones, January 11, 1944, and Caldwell to C. M. Goethe, January 21, 1944, AAS Records.

45. C. M. Goethe to E. V. Jones, April 6, 1944; Otis Caldwell to Jones, April 10, 1944; Caldwell to Jones, October 27, 1944, AAS Records; report of the Research Committee, *JAAS* 23 (February 1953): 93.

46. E. C. L. Miller to J. Shelton Horsley, November 17, 1931, VAS records.

47. J. Shelton Horsley's master letter to potential contributors to the research endowment fund, June 15, 1937; records of the endowment fund for 1937, VAS Records.

48. George D. Palmer, "Scientific Research, the Hope of the South," *Science*, July 5, 1940, p. 5.

49. President's Message, *BGAS* 1 (December 1943): 2.

50. E. C. L. Miller to J. Shelton Horsley, November 17, 1931, VAS Records.

51. Proceedings of the AAS, March 13–14, 1931, *JAAS* 3 (1932): 14–16.

52. Emmett B. Carmichael to J. F Duggar, July 20, 1932, AAS Records.

53. Report of the Industrial Committee, *PVAS*, 1930–31, p. 11.

54. VAS Council meeting, May 8, 1932, *PVAS*, 1929–30, p. 9.

55. Report of the Industrial Committee, *PVAS*, 1930–31, p. 11.

56. *Transactions of the Illinois Academy of Science* 31 (June 1939): 280 and 41 (May 1948): 126; *PIAS* 46 (1936): 11; 47 (1937): 10, 17; and 58 (1948): 10.

6. Southern Scientists and the Ideal of Service

1. Reingold, "Definitions and Speculations," 35.

2. For a broad overview, see Thomas L. Haskell, *The Authority of Experts: Studies in History and Theory* (Bloomington: Indiana University Press, 1984); Peter J. Kuznick, *Beyond the Laboratory: Scientists as Political Activists in 1930s*

America (Chicago: University of Chicago Press, 1987), 13–14. Kuznick's focus is primarily on scientists as political activists in the turbulent 1930s; however, the basis of their authority can be applied as well to activism on other fronts, including the environment.

3. McGill, "Tennessee Academy of Science," 8.

4. Programs of the TAS, *TTAS* 2 (1917): 9–12; Joseph Hyde Pratt, "The Southern Appalachian Forest Reserve," *JEMSS* 21 (December 1905): 156–64.

5. Pratt, "Southern Appalachian Forest Reserve," 156–64; Pratt, "The Occurrence and Utilization of Certain Mineral Resources of the Southern States," *JEMSS* 30 (June 1914): 1–25; (August 1914): 90–115; Pratt, "Conservation and Utilization," 1–25.

6. Proceedings of the NCAS, April 29–30, 1921, *JEMSS* 37 (December 1921): 6; proceedings of the NCAS, May 4–5, 1923, ibid. 39 (August 1923): 2, 8.

7. Proceedings of the NCAS, April 27–28, 1928, *JEMSS* 44 (September 1928): 7; proceedings of the NCAS, May 3–4, 1935, ibid. 51 (December 1935): 200.

8. Proceedings of the NCAS, May 7–8, 1937, *JEMSS* 53 (December 1937): 211–14.

9. Proceedings of the NCAS, May 6–7, 1938, *JEMSS* 54 (December 1938): 173; proceedings of the NCAS, May 5–6, 1939, ibid. 55 (December 1939): 222–25; proceedings of the NCAS, April 25–26, 1941, ibid. 57 (December 1941): 178–81.

10. Proceedings of the NCAS, May 8–9, 1953, *JEMSS* 69 (December 1953): 67–69; proceedings of the NCAS, May 7–8, 1954, ibid. 70 (December 1954): 109.

11. Report of the Secretary, *PVAS*, 1925–26, p. 4; minutes of the VAS meeting, May 7–8, 1926, ibid., 10.

12. Staggers, "History of the Virginia Academy of Science," 69; minutes of the VAS meeting, May 9–11, 1929, *PVAS*, 1929–30, p. 9; proceedings of the NCAS, May 4–5, 1923, *JEMSS* 39 (August 1923): 2, 8.

13. Minutes of the VAS meeting, May 9–10, 1930, *PVAS*, 1930–31, pp. 11–12.

14. Charles, "History of the Georgia Academy of Science," 128; minutes of the organizational meeting of the GAS, March 25, 1922, GAS Records.

15. Proceedings of the NCAS, April 27–28, 1917, *JEMSS* 33 (November 1917): 92; proceedings of the NCAS, May 4–5, 1923, ibid. 39 (August 1923): 1, 5–6; *Raleigh News and Observer*, December 3, 1922; proceedings of the NCAS, May 5–6, 1939, *JEMSS* 55 (December 1939): 220.

16. Proceedings of the NCAS, May 4–5, 1923, *JEMSS* 39 (August 1923): 1, 5–6; C. M. Heck to H. R. Totten, May 17, 1923, H. R. Totten Papers, Southern Historical Collection, University of North Carolina, Chapel Hill.

17. Report of High School Science Committee, proceedings of NCAS, May 2–3, 1924, *JEMSS* 40 (December 1924): 96–97.

18. Proceedings of the NCAS, May 6–7, 1927, *JEMSS* 43 (December 1927): 1.

19. Proceedings of the NCAS, May 8–9, 1931, *JEMSS* 47 (January 1932): 6.

20. Proceedings of the NCAS, May 7–8, 1937, *JEMSS* 53 (December 1937): 210; proceedings of the NCAS, April 25–26, 1941, ibid. 57 (December 1941): 176.

21. Proceedings of the TAS, November 27, 1925, *JTAS* 1 (January 1926): 9.

22. For a detailed study of the passage of the Butler Bill and the Scopes trial, see

Ray Ginger, *Six Days or Forever? Tennessee v. John Thomas Scopes* (Chicago: Quadrangle Books, 1969).

23. Staggers, "History of the Virginia Academy of Science," 63; E. C. L. Miller to Ivey F. Lewis, April 23 and April 28, 1924, VAS Records.
24. Henry Smith to Ivey F. Lewis, April 23, 1924, VAS Records.
25. Staggers, "History of the Virginia Academy of Science," 64.
26. Proceedings of the NCAS, April 30–May 1, 1926, *JEMSS* 42 (October 1926): 7. For a complete discussion of the controversy surrounding the theory of evolution in North Carolina, see Willard B. Gatewood, Jr., *Preachers, Pedagogues, and Politicians: The Evolution Controversy in North Carolina, 1920–1927* (Chapel Hill: University of North Carolina Press, 1966).
27. Proceedings of the TAS, November 29, 1929, *JTAS* 5 (January 1930): 21, 29–30; proceedings of the TAS, April 26–27, 1940, ibid. 16 (April 1941): 248.
28. Report of the Councilor to the AAAS, *JAAS* 4 (1933): 10.
29. John R. Sampey to Russell S. Poor, October 24, 1932, and Sampey to Emmett B. Carmichael, December 20, 1932, AAS Records.
30. John R. Sampey to Emmett B. Carmichael, January 7, 1933, and Sampey to Walter E. Jackson, January 21, 1933, AAS Records.
31. Proceedings of the AAS, March 10–11, 1933, and report of the Junior Academy Meeting, *JAAS* 5 (1934): 3, 7; Cantrell, Bailey, and Barker, *History of the Alabama Academy of Science*, 49–50.
32. Cantrell, Bailey, and Barker, *History of the Alabama Academy of Science*, 49–55.
33. Report of the Secretary, *PVAS*, 1925–26, p. 3.
34. Minutes of VAS meeting, May 9–10, 1930, *PVAS*, 1929–30, p. 13.
35. VAS Council meeting, April 24, 1931, *PVAS*, 1930–31, p. 17; report of the Committee on Junior Membership, ibid., 1931–32, pp. 16–18; minutes of VAS meeting, May 5–7, 1938, ibid., 1937–38, p. 20; proceedings of the VAS, May 2–4, 1940, *VJS* 1 (November 1940): 187; proceedings of the VAS, May 1–3, 1941, ibid. 2 (October 1941): 149.
36. Proceedings of the NCAS, April 30–May 1, 1920, *JEMSS* 36 (September 1920): 8; report of the executive committee, ibid. 37 (December 1921): 6.

7. A Dream Realized?
Advent of Research Facilities in the South

1. James C. Cobb, *The Selling of the South: The Southern Crusade for Industrial Development, 1936–1980* (Baton Rouge: Louisiana State University Press, 1982), 5–34 passim.
2. Proceedings of the TAS, November 29–30, 1912, *TTAS* 1 (1914): 18.
3. A. E. Parkins, "The Tennessee Valley Project—Facts and Fancies," *JTAS* 8 (October 1933): 345; proceedings of the TAS, December 1–2, 1933, ibid. 9 (January 1934): 67.
4. Parkins, "Tennessee Valley Project," 345–46.
5. Ibid., 352–53.
6. James C. Cobb, *Selling of the South.* especially 267–68.

7. George Fertig, "The Development of Scientific Research in the South," *Science*, June 10, 1932, pp. 595–99.

8. Ibid., 599.

9. Russell S. Poor, "The South's Position in the Mineral Industry," *JAAS* 7 (1935): 15.

10. Palmer, "Scientific Research," 1, 3.

11. Ibid., 3.

12. Ibid., 5.

13. Cattell, "Local Branches," 576–78.

14. J. L. Brakefield to Russell S. Poor, April 2, 1935, AAS Records; proceedings of the AAS, April 12–13, 1935, *JAAS* 7 (1935): 13; proceedings of the NCAS, May 3–4, 1935, *JEMSS* 51 (December 1935): 199–200; proceedings of the VAS, May 3–4, 1935, *PVAS*, 1934–35, p. 16.

15. George Palmer to George Boyd, March 23, 1940, AAS Records; meeting of AAS executive committee, March 29, 1940, *JAAS* 12 (June 1940): 6.

16. George Palmer to John Temple Graves, April 6, 1940; Harry Jennings to Palmer, September 5, 1940; Palmer to William M. Hinds, April 6, 1940; Palmer to E. D. McCluskey, April 6, 1940, records of the Southern Association of Science and Industry, Special Collections, Amelia Gayle Gorgas Library, University of Alabama, Tuscaloosa. See also *Birmingham News*, March 30, 1940, and *Birmingham Post*, March 30, 1940.

17. George Palmer to Wortley F. Rudd, July 10, 1940; Palmer to Leslie A. Sandholzer, July 17, 1940; Palmer to C. M. Farmer, July 17, 1940, AAS Records.

18. George Palmer to Wortley F. Rudd, July 10, 1940, AAS Records.

19. George Palmer to E. V. Jones, March 13, 1941; Palmer to Wortley Rudd, July 10, 1940; Palmer to C. M. Farmer, September 27, 1940; Palmer to E. V. Jones, October 4, 1940, AAS Records.

20. *Tuscaloosa News*, October 11, 1940.

21. George Palmer to E. V. Jones, October 15, 1940; form letter written by Palmer, February 8, 1941, AAS Records.

22. P. H. Yancey to Palmer, October 29, 1940, SASI Records.

23. Palmer to M. J. Funchess, January 7, 1941, SASI Records.

24. Palmer to Milton H. Fies, February 27, 1941; Palmer to Fies, March 5, 1941, SASI Records.

25. *Mobile Register*, March 23, 1941; a brief account of the meeting is contained in *JAAS* 13 (July 1941): 18; a copy of the printed program of this Mobile meeting is in the VAS Records.

26. *Mobile Register*, March 21, 1941.

27. Ibid.

28. Ibid.; *Mobile Press*, March 21, 1941.

29. Account of the meeting, *JAAS* 13 (July 1941): 18; George Palmer to E. V. Jones, April 12, 1941, AAS Records.

30. George Palmer, undated memorandum, SASI Records.

31. Form letter from Palmer to all those who attended the Mobile meeting, October 11, 1941, AAS Records.

32. *Atlanta Constitution*, April 3, 1942.

33. "A Review of Science-Industry Cooperation in the South during the Past Decade," *Journal of Southern Research* 1 (October 1949): 20–23; William

Pruett, "The Southern Association of Science and Industry," *Manufacturers' Record*, July 1957, pp. 49–56.

34. Irving E. Gray to Palmer, July 3, 1941, SASI Records.
35. E. C. L. Miller to Marcellus Stow, October 21, 1943, VAS Records.
36. George H. Boyd to George Palmer, June 26, 1945, SASI Records.
37. See, for instance, letters to George Palmer from B. W. Wells, professor of botany at North Carolina State College, February 17, 1947, and from Beatrice Nevins, professor of biology, Georgia State Woman's College, February 19, 1947, SASI Records.
38. H. M. Pace to George Palmer, September 3, 1947, SASI Records.
39. H. McKinley Conway to George Palmer, February 2, 1949, SASI Records.
40. H. McKinley Conway, editorial, *Journal of Southeastern Research* 1 (January 1949): 4; Paul W. Chapman, memo to all SASI officers, August 16, 1950, SASI Records.
41. Paul W. Chapman, memo to all SASI officers, August 16, 1950; copy of budget report for SASI for 1952–53 submitted to Office of Internal Revenue; McKinley Conway to George Palmer, January 15, 1955; Conway to Lloyd Bird, October 10, 1955; Conway to Palmer, May 16, 1958; John D. Wise to Palmer, April 1, 1959, SASI Records.
42. *Anniston Star*, October 13, 1940; *Tuscaloosa News*, October 11, 1940; *Birmingham Age-Herald*, October 11, 1940.
43. John Temple Graves, *History of Southern Research Institute* (Birmingham, Ala.: Birmingham Publishing, 1955), 11–16.
44. Carleton Ball to George Palmer, July 30, 1941; Dean Lee Bidgood to Palmer, November 7, 1941, SASI Records.
45. George Palmer to John R. McLure, November 8, 1941; Palmer to Thomas W. Martin, September 12, 1941, SASI Records.
46. Graves, *History of Southern Research Institute*, 34–35.
47. Ibid., 36–69 passim.
48. Mary Michelle Watkins and James A. Ruffner, eds., *Research Centers Directory*, 8th ed. (Detroit: Gale Research, 1983), 841.
49. Report on the Institute for Scientific Research, *PVAS*, 1944–45, p. 18; report on the Institute for Scientific Research, ibid., 1945–46, pp. 41–43; "The Virginia Institute for Scientific Research," undated memorandum in VAS Records.
50. Staggers and Flory, "History of the Virginia Academy of Science," 8; report for the Virginia Institute for Scientific Research, *PVAS*, 1948–49, p. 27; report for the Virginia Institute for Scientific Research, *VJS* 4 (September 1953): 198–99; Watkins and Ruffner, *Research Centers Directory*, 855; "The South's Big Bet on Technology," *Fortune*, March 1952, p. 92.
51. Walter M. Scott, editorial, *Journal of Southern Research* 1 (July 1949): 1; report on Institute of Industrial Research in Louisville, ibid., 24; interview with James W. Mullen, Jr., printed in *Journal of Southeastern Research* 1 (April 1949): 26–27.
52. "South's Big Bet on Technology," 92.
53. Scott, editorial, 1.
54. Report on the first Southwide Research Conference, *Journal of Southeastern Research* 1 (April 1949): 17–21.

55. James C. Cobb, *Selling of the South,* 175.
56. Ibid., 171–75.
57. Ibid., 175.
58. Cameron Fincher, *Research in the South: An Appraisal of Current Efforts,* Research Paper no. 5 (Atlanta: Georgia State College, 1964), 15–16.
59. Ibid., 16–17.

8. State Academies of Science in the Postwar World: Searching for an Identity

1. Address of the President-Elect, *VJS* 1 (November 1940): 181.
2. Report of the Secretary, VJS 2 (1941): 138.
3. Report of the Long Range Planning Committee, *VJS* 2 (1941): 158–59.
4. E. V. Jones, "Some Challenges Facing the Alabama Academy of Science," *JAAS* 16 (September 1944): 19.
5. Ibid., 22–23.
6. W. B. Redmond, Presidential Address, *BGAS* 5 (December 1947): 1–2; see also reports of membership committees and membership lists, ibid. 6 (December 1948) and 7 (December 1949).
7. Membership list, *BGAS* 8 (December 1950): 2.
8. E. Ruffin Jones, "Evolution of the Academy of Science," 230; report of the annual meeting, SCAS, *BSCAS* 43 (1981): 126.
9. R. T. Lagemann, Presidential Address, *BGAS* 9 (April 1951): 2.
10. AAS executive committee meeting, May 5, 1949, *JAAS* 21 (February 1951): 64.
11. See the reports of the Long Range Planning Committee and the annual programs throughout the 1950s and the 1960s, as printed in the *JAAS.* For an account of Von Braun's address, see *JAAS* 36 (July 1965).
12. Proceedings of the annual meeting of the NCAS, *JEMSS* 91 (Summer 1975): 35. Figures for the VAS are contained in Staggers and Florry, "History of the Virginia Academy of Science."
13. Proceedings of the annual meeting of the SCAS, *BSCAS* 37 (1975): 21 and 38 (1976): 11; Bennett, "First Half-Century," 32–33.
14. E. V. Jones, "Some Challenges," 21; E. C. L. Miller to Jones, January 13, 1944, VAS Records.
15. E. C. L. Miller to E. V. Jones, January 26, 1944, VAS Records.
16. Sidney Negus to E. C. L. Miller, undated; Miller to Ernest V. Jones, January 26, 1944, VAS Records.
17. Lubow Margolena Hansen to Foley Smith, November 19, 1949, VAS Records. Hansen does not appear in *American Men of Science.*
18. E. V. Jones, "Some Challenges," 21; meeting of the AAS executive committee, April 29, 1955, *JAAS* 27 (December 1955): 112–13; proceedings of the NCAS, May 4–5, 1951, *JEMSS* 67 (December 1951): 163.
19. Margaret W. Rossiter, *Women Scientists in America: Struggles and Strategies to 1940* (Baltimore: Johns Hopkins University Press, 1982), xvii.
20. J. S. Holmes to J. B. Bullitt, March 14, 1942, William C. Coker Papers.

21. E. V. Jones, "Some Challenges," 23–24.
22. NCAS Treasurer's Report, *JEMSS* 88 (Fall 1982): 173; reports of the editor, ibid. 89 (Winter 1973): 228 and 94 (Summer 1978): 34.
23. Report of the Secretary, May 6, 1948, *PVAS*, 1947–48, p. 15.
24. Personal interview with Boyd Harshbarger, July 18, 1983, Blacksburg, Virginia.
25. Staggers and Flory, "History of the Virginia Academy of Science," 18–28; meeting of the VAS council, May 10, 1956, *VJS* 4 (September 1956): 218.
26. Staggers and Flory, "History of the Virginia Academy of Science," 28.
27. Perry C. Holt, "The *Virginia Journal of Science:* Its Status and Goals," *VJS* 24 (1973): 73–74.
28. Report of the editor, *VJS* 29 (Summer 1978): 27.
29. Reports of the Conservation Committee, NCAS, *JEMSS* 68 (December 1952): 119; 69 (December 1953): 70; 70 (December 1954): 110; 77 (November 1961): 94.
30. Report of the Conservation Committee, NCAS, *JEMSS* 80 (December 1964): 143; 81 (November 1965): 60; 83 (Fall 1967): 154.
31. Proceedings of the NCAS, *JEMSS* 90 (Fall 1974): 80; proceedings of the VAS, *VJS* 31 (Winter 1980): 74.

9. Educating the Next Generation: The Academies' Role

1. Report of the Committee on the Junior Academy, *JAAS* 5 (1934): 19; Cantrell, Bailey, and Barker, *History of the Alabama Academy of Science*, 49–50.
2. Cantrell, Bailey, and Barker, *History of the Alabama Academy of Science*, 55–56; Kuznick, *Beyond the Laboratory*, 13.
3. J. M. Robinson, "The Alabama Academy of Science in the Postwar Era," *JAAS* 18 (December 1946): 83.
4. Cantrell, Bailey, and Barker, *History of the Alabama Academy of Science*, 56–59.
5. Hubert J. Davis, "The Junior Academy Movement," *VJS* 2 (1941): 57.
6. Proceedings of the NCAS, April 27–28, 1973, *JEMSS* 89 (Winter 1973): 236.
7. Proceedings of the NCAS, May 8–9, 1931, *JEMSS* 47 (January 1932): 5.
8. Proceedings of the NCAS, May 3–4, 1935, *JEMSS* 51 (December 1935): 199.
9. Davis, "Junior Academy Movement," 62.
10. Margaret E. Patterson, "The National Science Talent Search: Its History and Accomplishments," *BGAS* 13 (January 1955): 6–7.
11. Proceedings of the NCAS, May 3–4, 1946, *JEMSS* 62 (December 1946): 118.
12. Harold A. Edgerton, Steuart Henderson Britt, and Helen M. Davis, "Is Your State Discovering Its Science Talent?" *Science Education* 28 (October 1944): 228–31.
13. Jacob W. Shapiro, "How Tennessee Science Teachers May Discover and Develop Science Talent," *JTAS* 21 (1946): 189–90; report of the Tennessee Science Talent Search, ibid. 25 (1950): 298–99.

14. Staggers and Flory, "History of the Virginia Academy of Science," 5; report of Virginia Science Talent Search Committee, *PVAS*, 1945–46, pp. 29–36.
15. Report of Virginia Science Talent Search Committee, *PVAS*, 1947–48, pp. 28–32; Staggers and Flory, "History of the Virginia Academy of Science, 5.
16. Cantrell, Bailey, and Barker, *History of the Alabama Academy of Science*, 60.
17. Ibid., 60–61; meeting of AAS executive committee, May 1, 1947, *JAAS* 19 (December 1947): 55.
18. Cantrell, Bailey, and Barker, *History of the Alabama Academy of Science*, 63–64.
19. Ibid., 64; report of the State Talent Search, *JAAS* 20 (December 1948): 96–97.
20. Cantrell, Bailey, and Barker, *History of the Alabama Academy of Science*, 65–67; a report of the Gorgas Scholarship Foundation is contained in each volume of the *JAAS*.
21. Cantrell, Bailey, and Barker, *History of the Alabama Academy of Science*, 69.
22. For evidence of this concern about the required participation of students in the science fairs, see the proceedings of the NCAS, May 8–9, 1964, *JEMSS* 80 (December 1964): 140; see also *BSCAS* 29 (1967): 7.
23. For the relationship between science and the federal government, see Dupree, *Science in the Federal Government*; J. Merton England, *A Patron for Pure Science: The National Science Foundation's Formative Years, 1945–57* (Washington, D.C.: National Science Foundation, 1982), especially 236–37; Dorothy Schaffter, *The National Science Foundation* (New York: Frederick A. Praeger, 1969), especially 96; John T. Wilson, *Academic Science, Higher Education, and the Federal Government, 1950–1983* (Chicago: University of Chicago Press, 1983), 1–14.
24. See reports of the treasurer and minutes of business meetings in *JAAS* as follows: 31 (July 1959): 44; 33 (July 1962): 170; 34 (July 1963): 105; 36 (July 1965): 138.
25. See reports of the treasurer and minutes of business meetings in *JEMSS* as follows: 75 (November 1959): 54; 76 (November 1960): 170; 78 (November 1962): 81; 79 (November 1963): 78; 80 (December 1964): 144; 82 (November 1966): 67; 86 (Winter 1970): 158–59.
26. See reports of the treasurer and minutes of business meetings in *VJS* as follows: 6 (September 1955): 206; 7 (September 1956): 225; 8 (September 1957): 234; 10 (September 1959): 215; 11 (September 1960): 153; 12 (April 1961): 55; 15 (September 1964): 246–47; 17 (September 1966): 239; 18 (July 1967): 118; 20 (Summer 1969): 93; 22 (Spring 1971): 62.
27. GAS grant proposal to NSF, n.d. [1959]; Harry C. Kelly to Joseph P. LaRocca, April 24, 1959; NSF news release, April 3, 1962; GAS grant proposal to NSF, n.d. [1962]; Paul A. Scherer to Lucille Burnett, March 24, 1962; Bowen C. Dees to William H. Jones, March 14, 1963, GAS Records.
28. Report of the committee on the visiting scientists program, *VJS* 18 (July 1967): 118.
29. NCAS Treasurer's report, *JEMSS* 86 (Winter 1970): 158–59; TAS Treasurer's reports, *JTAS* 36 (April 1961): 166 and 45 (April 1970): 45; AAS Treasurer's report, *JAAS* 41 (July 1970): 218.
30. John T. Wilson, *Academic Science*, 33–37.

31. Report of the special committee on legislative relations, April 25, 1957, minutebook of AAS, AAS Records.
32. TAS Treasurer's reports, *JTAS* 35 (April 1960): 144 and 55 (April 1980): 47.

10. Looking to the Future

1. Proceedings of the AAS, *JAAS* 60 (July 1989): 221; proceedings of the GAS, *GJS* 47 (1989): 146.
2. Information on presentations and abstracts compiled from *JAAS* 60 (July 1989); *GJS* 48 (1990); *JTAS* 65 (April 1990): vote in TAS on funding graduate students from ibid., 31.
3. Arthur Bevan, "Science on the March: A Modern State Academy of Science," *Scientific Monthly* 73 (October 1951): 258–59.
4. Report of Gorgas Scholarship Committee, *JAAS* 60 (July 1989): 217–19.
5. Report of the Visiting Scientist Program, *JTAS* 65 (April 1990): 30, 32.
6. Report of Tennessee Junior Academy of Science, *JTAS* 65 (April 1990): 31.
7. Ibid., 36.
8. Wintfred L. Smith, "Report of the Director of the Reelfoot Lake Program," *JTAS* 52 (January 1977): 1.
9. GAS Budget, *GJS* 46 (1988): 211.
10. TAS Budget, *JTAS* 65 (April 1990): 32.

Selected Bibliography

Published Sources

Armes, Ethel. *The Story of Coal and Iron in Alabama*. Birmingham, Ala.: Chamber of Commerce, 1910. Reprint. New York: Arno Press, 1973.

Bailey, William M. "The Beginning of the Illinois State Academy of Science." *Illinois Academy of Science Transactions* 43 (1950): 24–33.

Baker, C. L. *History of the Academy Conference, 1926–1952*. Washington, D.C.: American Association for the Advancement of Science, 1952.

Barnhart, John Hendley. "The First Hundred Years of the New York Academy of Sciences." *Scientific Monthly* 5 (November 1917): 463–75.

Baskerville, Charles. "Science and the People." *JEMSS* 20 (1904): 68–79.

Bates, Ralph S. *Scientific Societies in the United States*. 3d ed. Cambridge: MIT Press, 1965.

Battle, Kemp P. "Dr. Joseph Austin Holmes at the University of North Carolina." *JEMSS* 32 (April 1916): 20–23.

———. *History of the University of North Carolina*. 2 vols. Raleigh: Edwards & Broughton, 1907–12. Reprint. Spartanburg, S.C.: Reprint Company, 1974.

Bayley, W. S. "The Place of State Academies of Science among Scientific Organizations." *Science*, June 1, 1923, pp. 623–29.

Beer, John J., and W. David Lewis. "Aspects of the Professionalization of Science." *Daedalus* 92 (1963): 754–84.

Bennett, Harry J. "The First Half-Century of the Louisiana Academy of Science, 1927–1977." *PLAS* 45 (December 1982): 8–88.

Bevan, Arthur. "Science on the March: A Modern State Academy of Science." *Scientific Monthly* 73 (October 1951): 255–60.

Bledstein, Burton J. *The Culture of Professionalism: The Middle Class and the Development of Higher Education in America*. New York: W. W. Norton, 1976.

Boyd, Paul P. "The Future of the State Academy of Science." *Science*, June 11, 1920, pp. 575–80.

Brubacher, John S., and Willis Rudy. *Higher Education in Transition: An American History, 1636–1956.* 3d ed. New York: Harper & Row, 1976.

Bruce, Philip Alexander. *History of the University of Virginia, 1819–1919.* 5 vols. New York: Macmillan, 1921.

Bruce, Robert. *The Launching of Modern American Science, 1846–1876.* New York: Alfred A. Knopf, 1987.

———. "A Statistical Profile of American Scientists, 1846–1876." In *Nineteenth-Century American Science: A Reappraisal,* edited by George Daniels, 63–94. Evanston, Ill.: Northwestern University Press, 1972.

Buckner, Ellen B. "Outstanding Alabama Scientists: Profiles of Five Years' Recipients of the Wright A. Gardner Award." *JAAS* 60 (April 1989): 49–60.

Bursey, Maurice M. *Carolina Chemists: Sketches from Chapel Hill.* Chapel Hill: University of North Carolina, Department of Chemistry, 1982.

Butler, Amos W. "Early History of the Indiana Academy of Science." *PIAS* 33 (1923): 14–18.

Butts, Charles. "Memorial of Eugene Allen Smith." *BGSA* 39 (1927): 51–65.

Cantrell, Clyde H., Paul C. Bailey, and S. B. Barker, eds. *A History of the Alabama Academy of Science.* Auburn: Alabama Academy of Science, 1963.

Cattell, J. McKeen, ed. *American Men of Science.* Lancaster, Pa.: Science Press, 1906.

———. "Local Branches of the American Association for the Advancement of Science." *Science,* December 21, 1934, pp. 576–78.

Cobb, Collier. "The Garden, Field, and Forest of the Nation." *JEMSS* 23 (1907): 52–70.

Cobb, James C. *The Selling of the South: The Southern Crusade for Industrial Development, 1936–1980.* Baton Rouge: Louisiana State University Press, 1982.

Coker, W. C. "Science Teaching." *North Carolina High School Bulletin,* April 1910.

Conkin, Paul. *Gone with the Ivy: A Biography of Vanderbilt University.* Knoxville: University of Tennessee Press, 1985.

Dabney, Virginius. *Mr. Jefferson's University: A History.* Charlottesville: University Press of Virginia, 1981.

Dana, Edward S. "The American Journal of Science from 1818 to 1918." *American Journal of Science* 196 (1918): 1–44.

Daniels, George. *American Science in the Age of Jackson.* New York: Columbia University Press, 1968.

———, ed. *Nineteenth-Century American Science: A Reappraisal.* Evanston, Ill.: Northwestern University Press, 1972.

———. "The Process of Professionalization in American Science: The Emergent Period, 1820–1960." *Isis* 58 (1967): 151–66.

Davis, Hubert J. "The Junior Academy Movement." *VJS* 2 (1941): 57–62.

Directory and Handbook of the Association of Academies of Science, 1978–1979. Columbus: Ohio Academy of Science, 1979.

Dupree, A. Hunter. *Science in the Federal Government: A History of Policies and Activities to 1940.* Cambridge: Harvard University Press, Belknap Press, 1957.

Dyer, Thomas G. *The University of Georgia: A Bicentennial History, 1785–1985.* Athens: University of Georgia Press, 1985.

Eaton, Clement. *The Mind of the Old South.* Rev. ed. Baton Rouge: Louisiana State University Press, 1967.

Edgerton, Harold A., Steuart Henderson Britt, and Helen M. Davis. "Is Your State Discovering Its Science Talent?" *Science Education* 28 (October 1944): 228–31.

Emigh, Eugene D. "Southern Research and Education." *Journal of Southeastern Research* 1 (July 1949): 27–28.

England, J. Merton. *A Patron for Pure Science: The National Science Foundation's Formative Years, 1945–57.* Washington, D.C.: National Science Foundation, 1982.

Ewan, Joseph, and Nesta Ewan. "San Francisco as a Mecca for Nineteenth-Century Naturalists." In *A Century of Progress in the Natural Sciences, 1853–1953.* San Francisco: California Academy of Sciences, 1955.

Fertig, George. "The Development of Scientific Research in the South." *Science,* June 10, 1932, pp. 595–99.

Fincher, Cameron. *Research in the South: An Appraisal of Current Efforts.* Research Paper no. 5. Atlanta: Georgia State College, 1964.

Gardner, Wright A. "The Organization of the Alabama Academy of Science." *JAAS* 3 (1932): 7–9.

Gatewood, Willard B., Jr. *Preachers, Pedagogues, and Politicians: The Evolution Controversy in North Carolina, 1920–1927.* Chapel Hill: University of North Carolina Press, 1966.

Ginger, Ray. *Six Days or Forever? Tennessee v. John Thomas Scopes.* Chicago: Quadrangle Books, 1969.

Gordon, Charles Henry. "Science and Progress in the South." *TTAS* 1 (1914): 51–58.

Grantham, Dewey W. *Southern Progressivism: The Reconciliation of Progress and Tradition.* Knoxville: University of Tennessee Press, 1983.

Graves, John Temple. *History of Southern Research Institute.* Birmingham, Ala.: Birmingham Publishing, 1955.

Greene, John C. "American Science Comes of Age, 1780–1820." *Journal of American History* 55 (June 1968): 22–41.

———. *American Science in the Age of Jefferson.* Ames: Iowa State University Press, 1984.

Gunter, Gordon. "The Atmosphere Surrounding Science in Mississippi." *JMAS* 12 (1966): 29–44.

Guralnick, Stanley M. *Science and the Ante-Bellum American College.* Philadelphia: American Philosophical Society, 1975.

Hall, George Martin. "Memorial of Charles Henry Gordon, First President of the Tennessee Academy of Science." *JTAS* 10 (1935): 101–3.

Ham, L. B. "Early History: The Arkansas Academy of Science." *Proceedings of the Arkansas Academy of Science* 1 (1941): 3–6.

Harshbarger, W. A. "The Kansas Academy of Science." *TKAS* 29 (1918): 35–41.

Haskell, Thomas L. *The Authority of Experts: Studies in History and Theory.* Bloomington: Indiana University Press, 1984.

———. *The Emergence of Professional Social Science: The American Social Science Association and the Nineteenth-Century Crisis of Authority.* Urbana: University of Illinois Press, 1977.

Haygood, Tamara Miner. *Henry William Ravenel, 1814–1887: South Carolina Scientist in the Civil War Era.* University: University of Alabama Press, 1987.

Henderson, Archibald. *The Campus of the First State University.* Chapel Hill: University of North Carolina Press, 1949.

———. "The Elisha Mitchell Scientific Society: Its History and Achievements." *JEMSS* 50 (December 1934): 1–13.

Hindle, Brooke. *The Pursuit of Science in Revolutionary America, 1735–1789.* Chapel Hill: University of North Carolina Press, 1956.

Holt, Perry C. "The *Virginia Journal of Science*: Its Status and Goals." *VJS* 24 (1973): 73–74.

Howell, A. C. *The Kenan Professorships.* Chapel Hill: University of North Carolina Press, 1956.

Hoy. W. E., Jr. "History of the South Carolina Academy of Science." *BSCAS* 1 (1935): 2–3.

Johnson, Thomas Cary, Jr. *Scientific Interests in the Old South.* New York: D. Appleton-Century, 1936. Reprint. Wilmington, Del.: Scholarly Resources, 1973.

Jones, E. Ruffin, Jr. "The Evolution of the Academy of Science." *Quarterly Journal of the Florida Academy of Sciences* 24 (December 1961): 229–38.

Jones, E. V. "Some Challenges Facing the Alabama Academy of Science." *JAAS* 16 (September 1944): 19–24.

Jones, Walter B. "Eugene Allen Smith." *Science*, January 6, 1928, pp. 7–9.

Kevles, Daniel J. *The Physicists: The History of a Scientific Community in Modern America.* 1971. Reprint. Cambridge, Mass.: Harvard University Press, 1987.

Kevles, Daniel J., Jeffrey L. Sturchio, and P. Thomas Carroll. "The Sciences in America, circa 1800." *Science*, July 5, 1980, pp. 27–32.

Kiger, Joseph C. *American Learned Societies.* Washington, D.C.: Public Affairs Press, 1963.

Kohlstedt, Sally Gregory. *The Formation of the American Scientific Community: The American Association for the Advancement of Science, 1848–60.* Urbana: University of Illinois Press, 1976.

Kuritz, Hyman. "The Popularization of Science in Nineteenth-Century America." *History of Education Quarterly* 21 (Fall 1981): 259–74.

Kusner, J. H. "The Academy during 1936." *Proceedings of the Florida Academy of Sciences* 1 (1937): 1–2.

Kuznick, Peter J. *Beyond the Laboratory: Scientists as Political Activists in 1930s America.* Chicago: University of Chicago Press, 1987.

Lefler, Hugh T., and Albert Ray Newsome. *North Carolina: The History of a Southern State.* 3d rev. ed. Chapel Hill: University of North Carolina Press, 1973.

Lloyd, Stewart J. *Eugene Allen Smith: Alabama's Great Geologist.* New York: Newcomen Society in North America, 1954.

Lyon, Scott C. "Possibilities for a Biological Station at Reelfoot Lake." *JTAS* 1 (January 1926): 10–15.

McGee, W. J. "Fifty Years of American Science." *Atlantic Monthly* 82 (September 1898): 307–20.

McGill, J. T. "History of the Reelfoot Lake Biological Station." *JTAS* 8 (January 1933): 22–33.

———. "The Tennessee Academy of Science." *JTAS* 2 (1927): 7–10.

McMath, Robert C., Jr., Ronald H. Bayor, James E. Brittain, Lawrence Foster, August W. Giebelhaus, and Germaine M. Reed. *Engineering in the New South: Georgia Tech, 1885–1985.* Athens: University of Georgia Press, 1985.

Mendenhall, T. C. "The Relation of the Academy to the State and to the People of the State." *Science,* December 24, 1915, pp. 881–90.

Numbers, Ronald L., and Janet S. Numbers. "Science in the Old South: A Reappraisal." *Journal of Southern History* 48 (May 1982): 163–84.

Odum, Eugene Pleasants, ed. *A North Carolina Naturalist, H. H. Brimley: Selections from His Writings.* Chapel Hill: University of North Carolina Press, 1949.

Oleson, Alexandra, and John Voss, eds. *The Organization of Knowledge in Modern America, 1860–1920.* Baltimore: Johns Hopkins University Press, 1979.

Oleson, Alexandra, and Sanborn C. Brown, eds. *The Pursuit of Knowledge in the Early American Republic: American Scientific and Learned Societies from Colonial Times to the Civil War.* Baltimore: Johns Hopkins University Press, 1976.

Palmer, George D. "Scientific Research, the Hope of the South." *Science,* July 5, 1940, pp. 1–5.

Palmer, Thomas Waverly, comp. *A Register of the Officers and Students of the University of Alabama, 1831–1901.* Tuscaloosa: University of Alabama, 1901.

Parkins, A. E. "The Tennessee Valley Project—Facts and Fancies." *JTAS* 8 (October 1933): 345–57.

Patterson, Margaret E. "The National Science Talent Search: Its History and Accomplishments." *BGAS* 13 (January 1955): 6–7.

Poor, Russell S. "The South's Position in the Mineral Industry" (abstract). *JAAS* 7 (1935): 15.

Powell, William S., ed. *Dictionary of North Carolina Biography.* 2 vols. Chapel Hill: University of North Carolina Press, 1986.

Pratt, Joseph Hyde. "The Conservation and Utilization of Our Natural Resources." *JEMSS* 26 (1910): 1–25.

———. "Memorial Sketch of Dr. Joseph Austin Holmes." *JEMSS* 32 (April 1916): 1–11.

———. "The Occurrence and Utilization of Certain Mineral Resources of the Southern States." *JEMSS* 30 (June 1914): 1–25; (August 1914): 90–115.

———. "The Southern Appalachian Forest Reserve." *JEMSS* 21 (December 1905): 156–64.

Pruett, William. "The Southern Association of Science and Industry." *Manufacturers' Record,* July 1957, pp. 49–56.

Reingold, Nathan. "Definitions and Speculations: The Professionalization of Science in America in the Nineteenth Century." In *The Pursuit of Knowledge in the Early American Republic: American Scientific and Learned Societies from Colonial Times to the Civil War,* edited by Alexandra Oleson and Sanborn C. Brown, 33–69. Baltimore: Johns Hopkins University Press, 1976.

"A Review of Science-Industry Cooperation in the South during the Past Decade." *Journal of Southern Research* 1 (October 1949): 20–23.

Robinson, J. M. "The Alabama Academy of Science in the Postwar Era." *JAAS* 18 (December 1946): 83–85.

Rossiter, Margaret W. *Women Scientists in America: Struggles and Strategies to 1940.* Baltimore: Johns Hopkins University Press, 1982.

Rowland, Henry A. "A Plea for Pure Science." AAAS *Proceedings* 32 (1883): 105–26.

Rudolph, Frederick. *The American College and University: A History.* New York: Vintage Books, 1962.

Runk, B. F. D. "Ivey Foreman Lewis." *VJS* 5 (April 1954): 51–52.

Russell, Ernest E. "Status of the Academy of Science Today." *JMAS* 14 (1968): vii–x.

Schaffter, Dorothy. *The National Science Foundation.* New York: Frederick A. Praeger, 1969.

Schneer, Cecil J., ed. *Two Hundred Years of Geology in America.* Hanover, N.H.: University Press of New England, 1979.

Schoewe, W. H. "The Kansas Academy of Science—Past, Present, and Future." *TKAS* 41 (1938): 399–413.

Sellers, James B. *History of the University of Alabama.* Vol. 1, *1818–1902.* University: University of Alabama Press, 1953.

Shapiro, Jacob W. "How Tennessee Science Teachers May Discover and Develop Science Talent." *JTAS* 21 (1946): 189–90.

Smith, Eugene Allen. "Memorial of Eugene Woldemar Hilgard." *BGSA* 28 (1916): 40–55.

Smith, Wintfred L. "Report of the Director of the Reelfoot Lake Program." *JTAS* 52 (January 1977): 1.

Snyder, Carl. "America's Inferior Position in the Scientific World." *North American Review* 174 (January 1902): 59–72.

"The South's Big Bet on Technology." *Fortune,* March 1952, pp. 92–95.

Staggers, Harry Joseph. "A History of the Virginia Academy of Science, 1923–1945." *VJS* 19 (Winter 1968): 57–89.

Staggers, Harry Joseph, and Walter S. Flory. "A History of the Virginia Academy of Science, 1945–1973." *VJS* 24 (Winter 1973): 5–64.

Stephens, Lester D. *Joseph LeConte: Gentle Prophet of Evolution.* Baton Rouge: Louisiana State University Press, 1983.

Venable, Francis Preston. "The Elisha Mitchell Scientific Society—Historical Sketch, Growth and Development." *Carolina Chemist* 8 (May 1922): 10–11.

———. "Historical Sketch of the Elisha Mitchell Scientific Society." *JEMSS* 39 (April 1924): 117–22.

———. "Isotopes." *JEMSS* 37 (1921–22).

———. "Joseph Austin Holmes." *JEMSS* (April 1916): 16–23.

———. "The Mitchell Society." *Alumni Quarterly* 1 (October 1894): 39–41.

Veysey, Laurence R. *The Emergence of the American University.* Chicago: University of Chicago Press, 1965.

Watkins, Mary Michelle, and James A. Ruffner, eds. *Research Centers Directory.* 8th ed. Detroit: Gale Research, 1983.

Whitney, David D. "State Academies of Science." *Science,* December 5, 1919, pp. 517–18.

Wilcox, Harold E. "State Science Academies." *Journal of Southern Research* 3 (November–December 1951): 24–25.

Wilson, John T. *Academic Science, Higher Education, and the Federal Government, 1950–1983.* Chicago: University of Chicago Press, 1983.

Wilson, Louis Round. *The University of North Carolina, 1900–1930: The Making of a Modern University.* Chapel Hill: University of North Carolina Press, 1957.

Publications of Academies of Science and Research Institutes

Alabama Academy of Science. Proceedings, in *JAAS* 1 (1930) through 60 (1989).
Alabama Industrial and Scientific Society. Proceedings printed annually in *Proceedings of the Alabama Industrial and Scientific Society*, 1891–99.
Elisha Mitchell Scientific Society. Proceedings, in *JEMSS*, 1884–1983.
Georgia Academy of Science. Proceedings, in *BGAS*, 1943–82; and *GJS*, 1942–90.
Georgia State Horticultural Society. Proceedings, in *Proceedings of the Georgia State Horticultural Society*, 1876–98.
Illinois Academy of Science. Proceedings, 1939, 1948, in *Illinois Academy of Science Transactions* 31 (June 1939): 280; 41 (May 1948): 126.
Indiana Academy of Science. Proceedings, 1886–92, 1936–37, 1948, in *PIAS* 1 (1892): 46; (1936): 11; 47 (1937): 10, 17; 58 (1948): 10.
Journal of Southeastern Research 1 (April 1949) through (October 1949). Various editorials, interviews with southern scientists and research personnel, and reports of meetings.
Journal of Southern Research 1 (October 1949) through 8 (December 1956). Various editorials, interviews, and reports.
North Carolina Academy of Science. Proceedings, in *JEMSS*, 1904–82.
South Carolina Academy of Science. Proceedings, in *BSCAS*, 1935–82.
Tennessee Academy of Science. Proceedings, in *TTAS*, 1914 and 1917; and *JTAS*, 1926–89.
Virginia Academy of Science. Proceedings, in *PVAS*, 1923–24 to 1949–50; *VJS*, 1940–42 and 1950–89.

Manuscript Sources

Alabama Academy of Science. Records. Archives, Draughon Library, Auburn University, Auburn, Alabama.
Alabama Industrial and Scientific Society. Minutebook, membership lists, treasurers' reports. Special Collections, Amelia Gayle Gorgas Library, University of Alabama, Tuscaloosa.
Ashe, William Willard. Papers. Southern Historical Collection, University of North Carolina, Chapel Hill.
Brimley, Herbert Hutchinson. Papers. North Carolina State Archives, Raleigh.
Coker, Robert Ervin. Papers. Southern Historical Collection, University of North Carolina, Chapel Hill.
Coker, William C. Papers. Southern Historical Collection, University of North Carolina, Chapel Hill.
Elisha Mitchell Scientific Society. Records. In the possession of Maurice Bursey, Department of Chemistry, University of North Carolina, Chapel Hill.
Georgia Academy of Science. Records. Currently in the possession of George A. Rogers, Georgia Southern College, Statesboro.
Kilgore, Benjamin Wesley. Papers. North Carolina State Archives, Raleigh.
LeConte Scientific Society. Records. South Caroliniana Library, University of South Carolina, Columbia.
North Carolina Academy of Science. Records. Special Collections, J. Y. Joyner Library, East Carolina University, Greenville.

South Carolina Academy of Science. Records. South Caroliniana Library, University of South Carolina, Columbia.

Southern Association for Science and Industry. Records. Special Collections, Amelia Gayle Gorgas Library, University of Alabama, Tuscaloosa.

Totten, H. R. Papers. Southern Historical Collection, University of North Carolina, Chapel Hill.

Virginia Academy of Science. Records. Special Collections, Virginia Polytechnic Institute and State University, Blacksburg.

Other Sources

Alumni Quarterly. University of North Carolina. 1 (October 1894).

Anniston Star, October 13, 1940.

Atlanta Constitution, April 3, 1942.

Birmingham Age-Herald, October 11, 1940.

Birmingham News, March 15, 1900; March 30, 1940.

Birmingham Post, March 30, 1940.

Carroll, P. Thomas. "Academic Chemistry in America, 1876–1976: Diversification, Growth, and Change." Ph.D. diss., University of Pennsylvania, 1982.

Charles, Allan D. "A History of the Georgia Academy of Science." Master's thesis, Emory University, 1968.

Dyer, Thomas G. "Science in the Antebellum College: The University of Georgia, 1801–1860." Paper presented at the First Barnard-Millington Symposium on Science and Medicine in the South, University, Mississippi, March 25–27, 1982.

Harshbarger, Boyd. Personal interview with the author, July 18, 1983, Blacksburg, Virginia.

Kilgore, Benjamin Wesley. Biographical clippings file, North Carolina State University Archives, D. H. Hill Library, Raleigh.

"Little Doc's Ambulance." A pamphlet issued by the University of Alabama in commemoration of the restoration of Eugene Allen Smith's Studebaker wagon, n.d.

Millbrooke, Anne. "Science and Government in the Old South." Paper presented at the First Barnard-Millington Symposium on Science and Medicine in the South, University, Mississippi, March 25–27, 1982.

Mobile Press, March 21, 1941.

Mobile Register, March 21, 23, 1941.

Raleigh News and Observer, March 23, November 28, 1902; May 10, 1905; December 3, 1922.

Smith, Eugene Allen. Scrapbooks. Special Collections, Amelia Gayle Gorgas Library, University of Alabama, Tuscaloosa.

Staggers, Harry Joseph. "A History of the Virginia Academy of Science, 1923–1945." Master's thesis, College of William and Mary, 1967.

Stephens, Lester D. "Scientific Societies in the Old South." Paper presented at the First Barnard-Millington Symposium on Science and Medicine in the South, University, Mississippi, March 25–27, 1982.

Stone, Bruce Winchester. "The Role of the Learned Societies in the Growth of Scientific Boston, 1780–1848." Ph.D. diss., Boston University, 1974.

Tuscaloosa News, October 11, 1940.

Index

235

About the Author

Nancy Smith Midgette is Associate Professor of History, Elon College. She received her bachelor's and master's degrees from North Carolina State University and her doctorate from the University of Georgia.

Midgette picks a theme for each chapter and then explores how each state society lived up to that theme, casually mentioning, also, those that did not. In this approach, definitions are key.

Book really emphasizes N.C., Tenn. Al, Vir. academies at the expense of others. Like nat'l overviews of science, some regions are overlooked. She measures vitality according to her criteria of publishing, membership, etc. Tyranny of numbers. What constitutes vitality?